Analysis of Volatiles

Analysis of Volatiles

Methods and Applications

Proceedings
International Workshop
Würzburg, Federal Republic of Germany
September 28 – 30, 1983

Editor
Peter Schreier

Walter de Gruyter · Berlin · New York 1984

Editor

Peter Schreier, Dr. rer. nat.
Professor of Food Chemistry
University of Würzburg
Food Chemistry
Am Hubland
D-8700 Würzburg
Federal Republic of Germany

CIP-Kurztitelaufnahme der Deutschen Bibliothek

Analysis of volatiles: methods and applications;
proceedings, internat. workshop, Würzburg, Fed. Republic of Germany,
September 28 - 30, 1983 / ed. Peter Schreier. –
Berlin; New York: de Gruyter, 1984.
ISBN 3-11-009805-9

NE: Schreier, Peter [Hrsg.]

Library of Congress Cataloging in Publication Data

Main entry under title:

Analysis of volatiles.

    Includes bibliographies and indexes.
      1. Food--Analysis--Congresses. 2. Flavor--Congresses.
3. Food--Odor--Congresses. I. Schreier, Peter,
1942- . II. Title: Volatiles.
TX511.A53 1984   664'.07   84-1721
ISBN 3-11-009805-9

PREFACE

The progress in the area of analysis of volatiles is closely
related to the development of instrumental measurement techniques, and nowadays we are being inundated with new information about analytical techniques. Today's analyst must make
many decisions that influence chromatographic and analytical
results. He has to select the appropriate sample preparation,
and with regard to the separation of volatiles - carried out
generally by the most powerful process, gas chromatography -
choices must be made as to length, diameter, stationary phase,
film thickness of the column, selection of glass or fused silica etc. The analyst makes decisions concerning the separating
parameters, and others relating to sample injection and detection that are essentially sample- and aim-dependent. Of course,
the analyst can benefit from all new developments in the field
of analysis of volatiles, but that requires exchange of information and experience, and basic understanding of the inter-
relationships involved.

Therefore, an International Workshop was held at the Chemistry
Department of the University of Würzburg from September 28-30,
1983, to bring together analysts working in different fields
of analysis of volatiles, e.g. in analytical, food, environmental, and biological chemistry, but who all use very similar
chromatographic and analytical techniques. The possibilities
and limits of different methods were discussed, and new trends
and perspectives of various analytical and chromatographic
techniques demonstrated. This book contains the lectures presented during the Workshop, covering the following main topics:
sample preparation, analytical techniques, and applications.

Many individuals and authorities contributed to the success of
the Workshop. In particular, the Editor is grateful to the
members of his staff who acted as registrators, operators,
technicians, and in many other functions, and who helped to
create a friendly working atmosphere.

The Editor also thanks all contributors for their ready collab-
oration in the preparation of this book. Special thanks go to
the Publisher's staff for their expert guidance given at all
stages of publication, and to Mrs. S. Grimm for her secretarial
assistance.

December 1983   P. Schreier
                                 University of Würzburg
                                 Food Chemistry

ACKNOWLEDGEMENTS

For economic support received to accomplish the workshop we
want to express our gratitude to the following companies,

BRUKER GmbH, Karlsruhe

CARLO ERBA Instruments, Hofheim

CHROMPACK GmbH, Mülheim

DANI GmbH, Mainz

DIGILAB GmbH, München

FINNIGAN-MAT GmbH, München

FRITZ GmbH, Hofheim

ICT GmbH, Frankfurt

NICOLET GmbH, Offenbach

ORION Corp. (Seekamp GmbH), Achim

PERKIN-ELMER GmbH, Überlingen

SIEMENS AG, Karlsruhe

WGA Analysentechnik, Düsseldorf

CONTENTS

SAMPLE PREPARATION

PREREQUISITES FOR SAMPLE PREPARATION IN FLAVOUR ANALYSIS

Hiroshi Sugisawa

Department of Food Science, Kagawa University, Mikicho,
Kagawa-ken, 761-07, Japan

Introduction

In flavour analysis, the first step is to obtain from the
food of interest, a sufficient quantity of the volatiles in
order to permit a meaningful analysis. Consequently, sample
preparation is an important aspect in flavour analysis. Too
often, emphasis is placed on the fractionation and identification steps, while selection of a representative sample and
isolation of the volatile compounds from the food are not
always stressed. The latter two aspects receive less attention, because fractionation and identification are more interesting, especially in view of the sophisticated instruments
such as gas chromatograph , mass spectrometer and nuclear
magnetic resonance spectrometer available to the researcher.

Several isolation techniques have been developed for volatiles
representative of the true characteristic flavour of a food.
The method used depends on the structure of the food, e.g. an
oil, an aqueous solution, or a solid mass.

Distillation, extraction, adsorption and other procedures are
frequently used for the isolation of flavour constituents
from specific food materials. Because the method used to isolate the flavour compounds can have profound effects on the
resultant"aromagram", our knowledge of flavour is influenced
by the procedure used in sample preparation and isolation.
Therefore, volatiles should be taken when the sample of inte-

rest is at its optimum flavour development. Recently, the important and critical aspects of sample preparation have been discussed briefly by Sugisawa (1), but improvement in sample selection and volatile isolation is still needed.

Sample Selection

Identification and quantitation of the volatiles that constitute the characteristic aroma of a particular food product, require that the product possesses the characteristic flavour and has no objectionable off-flavour.

1. Sample selection of fresh fruits and vegetables

Sample selection in the case of some fruits, vegetables or other plant products should be determined by the stage of maturation or ripeness, or by some other appropriate index. Problems rise when amounts of the material have to be processed and often the desired commodity is available only for a few days or weeks each year.

It is generally known that green odorous compounds such as C_6 aldehydes and C_6 alcohols are produced by enzymatic reaction, which catalyzes the oxidative splitting of linolenic acid and linoleic acid. This enzyme reaction usually proceeds fairly quickly and is complete within a few minutes in the presence of oxygen (2).

a) Enzymatic reactions. During the disentegration of green tea leaves, variable amounts of trans-2-hexenal and cis-3-hexenol are produced (Fig. 1). The level of these compounds reached a maximum in 3 min during disentegration of tea leaves (2).

Dirinck et al.(3) showed that the evaluation of strawberry flavour quality is possible by measuring the total amounts of

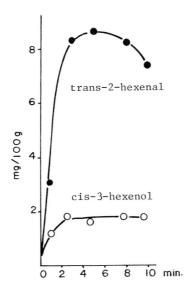

Fig. 1. Variation of C_6 compounds in
homogenates of fresh tea leaves(2).

volatiles through specific analysis of sulfur compounds. They
also found that the sulfur containing compounds such as dimethyl disulfide, methylthiol butyrate and methylthiol acetate
are important constituents in strawberry flavour. During maceration of some varieties in a blender and sniffing the
head space, they detected a rotten odour, which after
only a few minutes more resumed its fruity odor. The rotten
odor is caused by methanethiol, formed enzymatically from the
thiol esters during the maceration procedure of strawberries.

These studies demonstrated that the enzymatic reactions in the
macerated fruits or vegetables are rapid and can lead to noticeable changes in flavour quality. Precautions such as working
at low temperatures, in an inert gas atmosphere or in the absence of sunlight, may be required to avoid enzymatic reaction
and other chemical oxidation during the sample preparation.

6

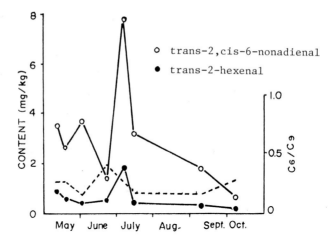

Fig. 2. Seasonal variation of a C$_9$ aldehyde and
        C$_6$ aldehyde in cucumber (2).

b) <u>Seasonal variation.</u>   Hatanaka (2) reported on the
seasonal changes of character impact flavours of fresh cucum-
ber. The amounts of trans-2, cis-6-nonadienal and trans-2-hex-
enal reached a maximum between June and July, but little change was noted during the rest of year (Fig. 2).

Ripening fruits are in a dynamic metabolic state, and short
term variation and wide fluctuation in the formation of individual volatiles are not uncommon.

Tressel and Jennings (4) investigated the changes in volatile
compounds of ripening bananas kept at 25°C. They observed that
acetate and butyrate ester formations were cyclical. The cycle
observed for the acetates was apparently out of phase with
that observed for the butyrates. Ethanol began to appear in
the postclimacteric ripening phase and increased thereafter.
These three compounds, acetates, butyrates and ethanol were
relatively minor components in the period of maximum desirable
flavour, but became major components 10-12 days later, a point
where the banana was over-ripe with an undesirable flavour.
Their result suggest that the most suitable period for desir-

able flavour may be lost after only a few days (3-4 days with
bananas).

The changes of fresh green odorous compounds, C_6 and C_9 aldehydes, in the bananas kept at room temperature were investigated by Hatanaka (2). Green bananas contained large quantities
of C_9 aldehyde (trans-2, cis-6-nonadienal), but the C_6 aldehyde (trans-2-hexenal) increased when the bananas turned
yellow. These results agree with the volatiles from unpeeled
banana reported by Tressl and Jennings (4).

The influence of maturity on the volatiles in Hamlin and Valencia orange peel was discussed by Lund and Dinsmore (5). A
remarkable variation was observed in the amounts of C_6 compounds such as hexanol, hexenol and hexenal in the Hamlin orange
(Fig. 3). Valencia orange peel aroma did not change as much
and the amounts of the C_6 compounds were also much lower.

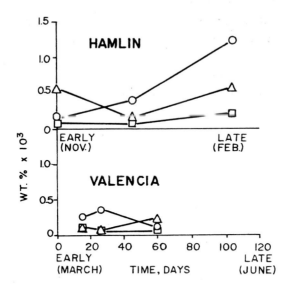

Fig. 3. Seasonal variation in the concentrations of cis-3-
hexen-1-ol (O), 1-hexanol (Δ), n-hexanal (□)
in orange peel aroma (5). *Copyright by Academic Press,
Inc. (1978).*

Another interesting observation was the variation of acetaldehyde in the juice. Acetaldehyde increased significantly in juice held at room temperature for 4 hrs, but not in juice that was pasteurized or kept at low temperature (4°C) (5).

2.   Sample selection in processed foods

Sample selection in processed foods that have been boiled, roasted, fried, or baked, becomes very difficult, especially if uniformity is to be realized. Some information on the sample before process is desirable.

a)   Complexity of samples. The volatiles in eggs and egg products as well as the influence of storage, feed source, and extraneous odors on these volatiles have been presented by Maga (6). Generally, feed and other feed additives have been thought to be the cause for off-flavours, but off-flavours were not found in all of the eggs. These kinds of results make sample selection all the more confusing and difficult.

The off-flavour described as fishy egg is caused by trimethylamine precursors in feeds made from fish meal, rapeseed meal and other soybean gums. With feeding herring meal, it was found that 2-7% of the resulting eggs smelled " fishy " and 17% had a " crabby " odor. Analysis of the off-flavoured eggs showed levels of trimethylamine greater than 10 times of that in normal eggs. Feed contaminated with wood chip or sawdust resulted in off-flavoured eggs that contained anisoles and other phenolic compounds (6).

b)   Influence of roasting time. Many products such as peanuts, almonds, coffee, and cocoa are roasted to achieve flavour changes acceptable for the products. A knowledge of the effect of roasting on the sensory attributes would therefore not only be desirable, but also necessary if quality control and flavour acceptance are to be maintained.

A review on the effect of roasting on the composition of coffee was presented by Reymond (7).

Vitzthum et al.(8) found that the ratios of methylfuran, butanone-2, and methanol were good indicators for evaluating the freshness of roasted coffee. An aged coffee showed a significant increase in methanol, but a decrease in butanone-2.

Ziegleder (9) used methylated pyrazines formed during roasting of cocoa to check the degree of roasting. The formation of pyrazines was dependent on the roasting time. Tetramethylpyrazine appeared in the cocoa beans very soon after the start of roasting and its content could be used as a general indicator of the quality of the cocoa. Trimethyl-pyrazine appeared more slowly and indicated the degree of roasting. Good correlations between flavour and the degree of roasting were found.

The influence of roasting time on the volatile fraction for light roasted, medium roasted and dark roasted peanuts was investigated by Buckholz et al.(10). The results indicated that the same volatiles were present in all samples, and that all roasting conditions differed only in the concentration of volatiles. A decrease in carbonyl compounds (pentanal) and a subsequent increase in pyrazine compounds (2-ethyl-6-methyl-pyrazine) were important to good peanut flavour.

c) Influence of milling. Cereal grains, depending on use, receive different degrees of milling (polishing).

Over 100 volatile compounds were identified in cooked rice by steam distillation, but none of the volatiles were found to have the characteristic odor of cooked rice (11). Tsugita et al.(12) found a significant difference in the volatiles between the cooked odor of 92% milled rice (milling yeild 92%, usual polishing) and those of the other three different degrees of milling (50-85%). Forty volatiles were identified in

10

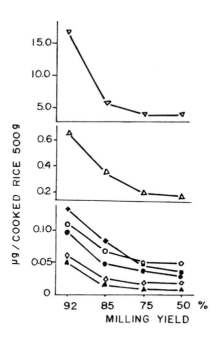

Fig. 4. Concentration of some typical volatile
components of cooked rice milled to
different degrees (modified from Ref.12).
∇ Hexanal;△ 2-pentylfuran;◆ 1-hexanol;
O 2-heptanone;● m-xylene;◇ 6-methyl-5-
hepten-2-one;▲ trans-2-hexenal.

*Copyright by the Agricultural Chemical Society of
Japan (1980).*

the cooked rice by headspace adsorption, and remarkable differences in volatile content and constituents were observed among
the samples. The study suggested that the outer layers of the
rice have an important role in the formation of cooked rice
volatiles. The relationship between volatiles in cooked rice
and the degree of milling is shown in Fig. 4.

d) Fried foods. Because of the growth in eating out at fast
food outlets, consumption of deep fried foods such as potatoes,
chicken, and fish has increased sharply in recent years. The
oil in deep frying is repeatedly used at elevated temperature
in the presence of oxygen. Under such conditions, both thermal

and oxidative decomposition of the oil occurs and volatile
and non-volatile decomposition products are formed. These
products have been extensively studied by Chang et al.(13).
They identified 220 compounds in the volatile fraction during
deep frying at 185°C. Thus, it is obvious that aromas of fried
food are quite complex, and the sample selection is difficult.

One solution to this problem is to have the characteristic
flavour determined by a panel test. An example of the latter
selection was discussed by Chang (14). He invited a number of
potato chip manufactures to bring their best samples to his
laboratory. These samples were then evaluated by a panel and
the sample selected as having the best flavour was subjected
to flavour analysis. A package of potato chips from the shelf
in a store may not be the ideal sample for flavour work.

Artifacts

The formation of an artifact during the isolation of volatiles
for instrumental analysis poses a serious problem. We can
never be absolutely certain that all of the compounds identified were present in the food sample. Artifacts arise from
chemical decomposition, enzymatic reaction or reaction among
the individual volatile components during the sample preparation. In some instances, artifacts may come from the anti-foa-
ming agent, laboratory distilled water, column and septum ble-
eds, vacuum grease, air pollutants in laboratory and from solvents (15).

a) Volatile air pollutants. Dirinck et al.(16) used a Tenax
GC adsorption method for isolation and concentration of volatiles from tomato, apple and strawberry. The procedure is relatively fast and yields volatiles quite representative of the
fruit. A commercial blender fitted with a three necked flange
type vessel equipped with a helium inlet and connected to an
adsorption column (Tenax GC 5g, 6 mm X 10 cm) was used.
Thirty three compounds from tomato, 32 compounds from apple

and 37 compounds from strawberry were identified. Before the
disentegration of the sample, the inside of blender was flushed with helium for 10 min, but contamination by volatile air
pollutants in the laboratory was a problem. Fourteen peaks
present in the laboratory air; such as benzene, toluene,
m-and p-xylene, were also found in the chromatogram. At present, this procedure is modified to avoid contamination by air
pollutants (3).

b) Phthalates and silicones. The pollution of phthalates can
become a serious problem in volatile work. Phthalates are present in plastic tubing, septums, and some solvents. Ishida et
al.(17) reported the presence of di-n-butyl phthalate and di-2
-ethylhexyl phthalate in solvents and in a few kinds of water
(Table 1).

Table 1. Concentrations of di-n-butyl phthalate (DBP)
and di-2-ethylhexyl phthalate (DEHP) in
organic solvents and waters (17).

Sample	DBP	DEHP	Other unknown peaks
City water*	2.04 ppb	–	–
Well water*	2.49 ppb	4.82 ppb	–
Tapp water(well water)	1.93 ppb	3.85 ppb	–
Ion-exchanged water	0.83 ppb	1.31 ppb	+
Benzene	0.17 ppm	1.96 ppm	+
Acetone	trace**	–	–
n-Hexane	–	–	–
Chloroform	–	–	–
Diethyl ether	–	43.6 ppb	+
Methanol	–	78.7 ppb	+
Ethanol	69.3 ppb	61.7 ppb	–
Light petroleum	–	–	–
Dichloromethane	–	–	–
Ethyl acetate	–	–	–
Acetonitrile	–	0.18 ppm	–

The organic solvent were purchased as GR materials
from Wako Chem. Co.

* From Tohoku University.
** Less than 10 ppb.   *Copyright by J. Chromatogr.(1980).*

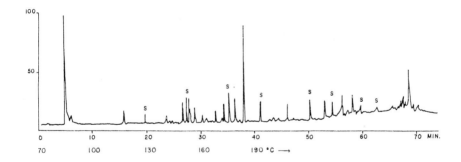

Fig. 5. GC of volatile components from callus tissues
of Perilla plant.

50 m X 0.25 mm, FFAP WCOT column. S: silicone
derivatives (characteristic mass spectral ion,
m/z 355, 295, 281, 221, 147, 73, 28).

Many kinds of silicone materials are used in the laboratory,
and contamination like that of phthalates causes similar problems. The volatile oil fraction from the callus tissues of
the Perilla plant(Akachirimen, Perilla frutescens Brit. var.
Crispa Decene f. purpurea Makino) was isolated by simultaneous
distillation adsorption method (18), and β-bisabolene, β-cham-
igrene, cuparene and 15 other sesquiterpene hydrocarbons were
found (19). However, eight silicone constituents were found in
this work (Fig. 5). The contaminants originated from a small
piece of plastic tubing used in the steam generator. Such artifacts can be avoided by an all glass system.

c) Non-volatile compounds. Lipids, waxes and other non-volatile compounds are sometimes extracted together with trace volatiles. These non-volatile compounds are deposited in the GC
injector or on the column and decompose by heating into various volatile components.

d) Other problems. Compositional changes of volatiles can be
caused by different distillation rates and thus the quantitative analysis will be affected.

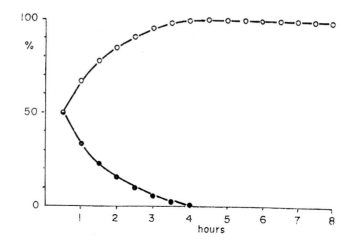

Fig. 6. Relative proportions of hydrocarbons (O) and
oxygenated compounds (●) in sequential fractions of Abies X arnoldiana oil(20).
Copyrigth by the American Chemical Society (1980).

Koedam et al.(20) followed the changes in the ratios of terpene hydrocarbons vs. oxygenated terpene compounds during continuous distillation of the volatile oil from plant material.
The first fraction (30 min distillation) contained approximately equal amounts of hydrocarbons and oxygenated compounds
(Fig. 6). With continuous distillation (1-4 hrs., European
Pharmacopoeia apparatus), the portion of the oxygenated compounds in the fractions decreased sharply and hydrocarbons increased substantially.

Therefore, the nature ot the sequential recovery of distillates
has to examined before a decision can be made as to which portion to utilize for the volatile studies. Distillation should
be done over a period of several hours to determine whether
volatile oils are distilled off. If the sample contains thermolabile compounds, artifacts may be produced by longer distillation periods.

Isomerization of components may occur during distillation, if

pH varies or if trace metals are present in the sample (20).
Artifacts created during the early stages of sample preparation cannot be eliminated at any later stage of compound identification.

Selective Characteristics of Extracting Solvents

Volatile extraction with solvents has been a common method in
flavour research and is usually used to extract volatiles from
a distillate obtained by steam distillation (see Table 2).
Extraction of large volumes of distillate invariably results
in large volumes of extractant and under these conditions,
extraction efficiency decreases, extraction time increases and
artifact formation is enhanced.

Extraction solvents are usually selected on the basis of their
extraction selectivity and boiling point. Ethyl ether is extensively used because its extraction efficiency is higher than
other solvents. Hydrocarbons have good extraction ability but
have a low efficiency. Pentane, isopentane, and fluorocarbons
have a low affinity for ethanol and thus are useful for isolation of volatiles from alcoholic beverages.

Table 2. Frequency distribution of isolation
         and concentration procedure.

	Paper No.	Distribution(%)
Distillation	67	25
Extraction	63	23
Adsorption	62	23
Headspace	55	21
Others	12	8

The data of frequency distribution between 1980-
1982 has been based on the numbers of paper appeared
in C. A. Search 93(1980)-98(1982).

Freon does not extract the C_1- C_4 alcohols, but its efficiency for alcohols above C_4 increases sharply (21). It is often used in solvent desorption of volatiles from porous polymers and charcoal.

Liquid carbon dioxide is an excellent solvent, has a low boiling point and has high selectivity towards esters, aldehydes, ketones and alcohols (22). Furthermore, liquid carbon dioxide extraction can be used to obtain a highly concentrated water-free extract from aqueous solutions. Supercritical carbon dioxide exhibits considerable solvent power when used above pressures of 70 bar and thus can be substituted for many solvents (23).

Summary

The prerequisites for sample preparation have been discussed in some detail. Steps used in sample preparation must not cause any loss of desired trace volatile components, nor change the volatile compositions nor introduce artifacts that would alter the characteristic flavour. In all procedures, care should be exercised to obtain isolates with the desired sensory properties. In some cases, such as fruits or vegetables, the source materials must be kept under specific conditions for proper development of the desired flavours. In each case, good judgment must be used in sample selection and preparation.

Acknowledgement

The co-operation of Dr. T. Tsugita, Department of Agricultural Chemistry, Tokyo University, in the literature survey is gratefully acknowledged.

References

1. Sugisawa, H.: Flavor Research, Recent Advances; Teranishi,

R., Flath,R.A., Sugisawa,H. ed.; p.15-51, Marcel Dekker, New York 1981.

2. Hatanaka, K.: Koryo, No. 117, 23-31 (1977).

3. Dirinck, P.J., de Pooter, H.L., Willaert, G.A., Schamp,N. M.: J. Agric. Food Chem. 29, 316-321 (1981).

4. Tressl, R.,, Jennings, W.G.: J. Agric. Food Chem. 20, 189-192 (1972).

5. Lund, E.D., Dinsmore, H.L.: Analysis of Food and Beverages; Charalambous, G. ed.; p.135-185, Academic Press, New York 1978.

6. Maga, J.A.: J. Agric. Food Chem. 30, 9-14 (1980).

7. Reymond, D.: Agriculture & Food Chemistry, Past, Present, Future; Teranishi, R. ed.; p.315-332, AVI, Westport 1978.

8. Vitzthum, O.G., Werkhoff, P.: Analysis of Food and Beverages; Charalambous, G. ed.; p.115-133, Academic Press, New York 1978.

9. Ziegleder, G.: Chem Abstr.96, 197983j (1982).

10. Buckholz, L.L.Jr., Daum, H., Stier, E., Trout, R.: J. Food Sci. 45, 547-554 (1980).

11. Yajima, I., Yanai, T., Nakamura, M., Sakakibara, H., Habu, T.: Agric. Biol. Chem. 42, 1229-1233 (1978).

12. Tsugita, T., Kurata, T., Kato, H.: Agric. Biol. Chem. 44, 835-840 (1980).

13. Chang, S.S., Peterson, R.J., Ho, C.T.: J. Am. Oil Chem.Soc. 55, 718-728 (1978).

14. Chang, S.S.: Food Technol. 27, 27-39 (1973).

15. Joen, I.J., Reineccius, G.A., Thomas, E.L.: J. Agric. Food Chem. 24, 433-434 (1976).

16. Dirinck, P., Schreyen, L., Schamp, N.M.: J. Agric. Food Chem. 25, 759-763 (1977).

17. Ishida, M., Suyama, K., Adachi, S.: J. Chromatogr. 189, 421-424 (1980).

18. Sugisawa, H., Hirose, T.: Flavour'81; Schreier, P. ed.; p. 287-299, Walter de Gruyter, Berlin 1981.

19. Nabeta, K., Sugisawa, H.: Plant Tissue Culture 1982; Fujiwara, A. ed.; p 289-290, Japan Association of Plant Tissue Culture, Tokyo 1982.

20. Koedam, A., Scheffer, J.J.C., Svendsen, A.B.: J. Agric. Food Chem. 28, 862-866 (1980).

21. Teranishi, R., Hornstein, I., Issenberg, P., Wick, E.L.: Flavor Research, Principles and Techniques, p. 37-76, Marcel Dekker, New York 1971.

22. Schultz, W.G., Randall, J.M.: Food Technol. 24, 1282-1286
    (1970).

23. Stahl, E., Schilz, W., Schütz, E., Willing, E.: Angew.
    Chem. Int. Ed. Eng. 17, 731-738 (1978).

COMPARISON OF METHODS FOR THE ISOLATION OF VOLATILE COMPOUNDS FROM AQUEOUS MODEL SYSTEMS

M.M. Leahy and G.A. Reineccius
Department of Food Science and Nutrition
University of Minnesota
St. Paul, Minnesota 55108, U.S.A.

Introduction

Over the past several years, there have been some excellent
studies evaluating selected methods for the isolation of volatiles from foods or model systems. Most of these studies have
been discussed in current review articles (1-5). A limitation
of some of these studies is that relatively high levels
(> 1 ppm) of volatiles were used in the work. While researchers
were once concerned with the problems of isolating volatiles
from foods at the ppm levels, they now are commonly working
at the low ppb or even ppt levels. Methods suitable for the
determination of volatiles in the ppm range often are not
suitable for trace analysis studies. A second limitation of
previous work is that generally each researcher chose a unique
set of compounds for method evaluation. Therefore, comparisons
between studies are often difficult.

The purpose of this study was to evaluate the qualitative and
quantitative performance of some of the commonly used methods
for the isolation of volatiles from foods using the same set
of test compounds at ppb concentrations. The methods chosen
for study were direct headspace (split, on-column and splitless), headspace concentration (adsorbent trap and Freon reflux), Nickerson-Likens (atmospheric and reduced pressure) and
solvent extraction (batch and continuous extraction using pen-

tane , dichloromethane and diethyl ether).

Methodology

Test compounds used in this study included ethanol, propanol, butanol, octane, decane, ethyl propanoate, ethyl butanoate, ethyl pentanoate, 2-heptanone, acetophenone, benzyl acetate, methyl salicylate, L-carvone, ß-ionone, ethyl methyl phenyl glycidate, and isoeugenol. These compounds were chosen since they represent several classes of compounds and a wide range of solubilities and boiling points (see Table 1). A stock solution of all test compounds was initially made up in pure methanol. One mL (or less) of this stock solution was diluted with Glenwood distilled spring water (Glenwood Company) to 1 L to make the desired concentration for analysis.

All analyses except the headspace concentration technique using Tenax were performed on a Hewlett Packard Model 5880 Gas Chromatograph (GC). The headspace concentration method was performed using a Hewlett Packard Model 7671 Purge and Trap Sampler which was attached to a Hewlett Packard Model 5840 GC. A bonded phase OV-225 fused silica capillary column (25 m x 0.32 mm) was used in all analyses (J & W Scientific). This column was operated as follows in the H.P. 5880 GC: column head pressure = 20 psig; linear velocity = 75 cm/sec; flow rate = approx. 5 mL/min; carrier gas = hydrogen; temperature profile: initial temperature = 45°C, rate 1 = 15°C/min, final temperature 1 = 130°C, rate 2 = 5°C/min, final temperature 2 = 190°C, final time = 0 min. Operating conditions of the H.P. 5840 GC are outlined in the adsorbent trapping section.

Table 1 Physical properties of volatile compounds used in
         this study

Compound	Boiling Point	Solubility
Ethanol	78.5°C	Water soluble
Propanol	97.4°C	Water soluble
Butanol	117.25°C	Slightly soluble in water, soluble in alcohol and ether
Octane	125.6°C	Water insoluble
Decane	174.1°C	Water insoluble
Ethyl Propanoate	99°C	Soluble 60 parts water, soluble in alcohol and ether
Ethyl Butanoate	120°C	Soluble in 150 parts water, soluble in alcohol and ether
Ethyl Pentanoate	145°C	Insoluble in water, soluble in alcohol
2-Heptanone	151.5°C	Very slightly soluble in water, soluble in alcohol and ether
Acetophenone	202°C	Slightly soluble in water, soluble in alcohol and ether
Benzyl Acetate	213°C	Insoluble in water, soluble in alcohol and ether
Methyl Salicylate	220-224°C	Soluble in 1500 parts water, soluble in ether
L-Carvone	230-231°C	Insoluble in water, soluble in alcohol and ether
B-Ionone	126-128°C	Very slightly soluble in water, soluble in ether
Methyl Anthranilate	135.5°C	Slightly soluble in water, soluble in ether
Ethyl Methyl Phenyl Glycidate	272-275°C	Insoluble in water, soluble in ether
Isoeugenol	266°C	Slightly soluble in water, soluble in ether

Direct Headspace Method - Split Injection

Samples (50 mL) for headspace analysis (100 ppb, 1 ppm and
10 ppm) were measured into 130 mL serum bottles, covered with
a Teflon cap liner (Arthur H. Thomas Co.), a rubber septum
and then an aluminium cap. This final cap was crimped tightly
on the bottle top. The sample was placed in a 60°C water bath
for 1 hr prior to GC analysis. The GC was set up with a 2 mL
headspace sampling loop. Therefore, when a sample was to be
analyzed, the fill line of the sampling loop was forced through
the sample bottle septum. Pressure within the bottle forced
headspace vapors into the sampling system and filled the sampling loop. Changing the position of the six port valve then
routed GC carrier gas through the sample loop sweeping the
headspace into the GC for analysis. Approximately a 1:10 split
was used for this analysis.

Results are reported as ng of each compound per mL headspace
and GC peak area. Nanograms/mL headspace was calculated from
peak areas knowing GC response for each compound from the injection of a known calibration mixture (liquid sample).

Direct Headspace - Splitless Injection

Samples (50 ppb, 100 ppb, 1 ppm and 10 ppm) were prepared for
analysis as described above for split injection. However, the
sample (1 mL) was injected into the GC via gas-tight syringe.
A low injection was used (25 seconds to empty 1 mL into the
GC inlet). The GC was operated in the splitless mode with 0.75
min prepurge time.

Two approaches were used to improve chromatographic resolution
when this large volume of headspace gas was injected without
split. The first method used spiking of the sample with 5 to
25 µL of pure hexane. This was done to simulate the typical

splitless condensation and solvent effect that occurs in the GC of liquid injections. The second approach involved immersing about 20 cm of the GC column in liquid nitrogen until the inlet purge was activated. This effectively cold trapped the injection as a sharp band. Quantitation was accomplished as described for split headspace analysis.

Direct Headspace - On-column Injection

Samples were prepared as previously described for the splitless headspace method except the sample was equilibrated for 15 or 60 min in the water bath. Sampling of the headspace was accomplished by first inserting an 18 gauge needle through the bottle septum and then threading the fused silica syringe needle through the metal needle. A 5 mL gas tight syringe was used for sampling (2 mL gas volume). A slow injection was used along with column cold trapping as previously described.

Headspace Concentration - Freon Reflux

The Freon reflux method used in this study is basically that of Jennings (6). Approximately 45 ml of Freon 12 (difluorodichloromethane, bpt. -30°C) was added to the precooled distillation pot and then the distillation column and dry ice/ethanol cold trap were attached. Pure nitrogen gas was bubbled through 150 mL of sample (50 ppb) at a rate of 100 mL/min. This purge gas was then fed into the refluxing Freon via the second neck of the distillation pot.

Purge times of 1 and 3 hr at room temperature and 60°C were used for sample collection. When the chosen collection time had elapsed, 2 mL of diethyl ether containing 1 µL 2-octanone as internal standard was added to the distillation pot. The condenser and distillation column were removed and the Freon was

permitted to evaporate. The remaining ether was then transferred to a 1/2 dram vial and the ether was evaporated under
a gentle stream of nitrogen to approx. 25 μL. Three μL of this
concentrate was then injected onto the GC for analysis. The
GC was operated with a 1:10 inlet split.

The internal standard method was used in quantification. Recoveries are based on the amount of each compound recovered
in Freon vs. the total amount present in the sample being
purged.

Headspace Concentration - Adsorbent Trapping

A Hewlett Packard Purge and Trap sample collection system was
used in this study. Ten mL of sample (50 ppb) were placed in
a 50 mL culture tube and then it was purged with helium
(120 mL/min). The purged volatiles were passed through an adsorbent trap (10.5 cm length and 4 mm internal diameter) containing Tenax. After the desired purge time (15,30, or 60 min),
the adsorbent trap was backflushed into the GC with helium
while the trap was heated for 3 min (180°C).

The GC was operated in the split mode (1:10) and the GC column
was cold trapped while the volatiles were desorbed from the adsorbent. The GC was operated under the following conditions:
column head pressure = 16 psig; carrier gas = helium; carrier
gas velocity = 40 cm/sec; oven temperature profile: temp. 1 =
45°C, time 1 = 6,5 min, Rate 1 = 20°C/min, Rate 2 (run time 9
min) = 5°C/min, Rate 3 (run time 10 min) = 11°C and final temp.
= 190°C.

Solvent Extraction

Solvent extraction was accomplished using both batch (separatory funnels) and continuous extraction processes. The batch processes were carried out using diethyl ether (Baker, ACS Grade), dichloromethane and pentane (MCB, Omnisolv, glass distilled) as extracting solvents. The diethyl ether was distilled prior to use while the other two solvents were used as received. Only dichloromethane and diethyl ether were used in the continuous extraction apparatuses.

All methods were evaluated for the extraction of the 50 ppb test mixture only. Five replicates were performed on each method of extraction. Quantification was based on the internal standard method.

Batch Extraction

Five hundred mL of test solution (50 ppb) was placed in a separatory funnel. This mixture was then extracted initially with 75 mL of solvent followed by four 50 mL extractions. Each extraction was shaken 2 mins. The solvent extracts were pooled, dried with anhydrous $MgSO_4$, internal standard added (2-octanone), filtered and then concentrated by gently evaporating the solvent over a steam bath under a gentle stream of nitrogen gas. The extract was evaporated to approx. 100 µL for GC analysis.

Continuous Extraction

Continuous extraction was carried out using a lighter-than-water (Ace Glass) for diethyl ether and a heavier-than-water extractor for the dichloromethane solvent. The diethyl ether extractor used 550 mL of test solution (50 ppb) and 100 mL of solvent. The ether was heated with a heating mantle such that a reflux rate of 120 drops/min was maintained. The extractor was fitted with a double walled Davies condenser which was

cooled by recirculating ice water. This extraction was carried
out for 2 and 4 hr. At the end of extraction, the ether at
the top of the water was decanted into the ether distillation
pot, dried with anhydrous $MgSO_4$, internal standard added (2-
octanone), filtered and then concentrated under a stream of
nitrogen over a steam bath. The solvent extract was evaporated
to 100 μL for GC analysis.

The dichloromethane extractor was smaller in volume accommo-
dating 200 mL of sample and 60 mL of dichloromethane. The same
procedures were followed for this extraction as described above
for the continuous ether extraction.

Simultaneous Distillation/Extraction

A Nickerson-Likens extractor as modified by Schultz et al.
(7) was used in this study. A dry ice/ethanol cooled Dewar
flask was added to the vacuum port in order to help retain
solvent in the system when vacuum was applied. Even though the
distillation head was cooled by circulating ice water, the solvent was slowly lost from the system without this additional
condenser.

One L of sample (50 ppb) was added to a 2 L flask while approx.
25 mL diethyl ether was added to the 50 mL solvent flask. The
distillation head was attached and then both flasks were heated
to boiling. The distillation was carried out for 1,2 and 4 hr
at atmospheric pressure. Only 2 and 4 hr extractions were used
for the vacuum (400 mm absolute pressure) distillations.

At the end of extraction the ether was decanted back into the
solvent pot, dried with anhydrous $MgSO_4$, internal standard
added (2-octanone), filtered and then concentrated as has been
described for the other methods.

Results and Discussion

Headspace Methods

Headspace sampling is an attractive method for analyzing the
volatiles in foods. It is simple, rapid and measures the odors
in the proportions typically presented to the human nose. The
primary disadvantage is that the method lacks sensitivity (8,
9). Typically only the most volatile and abundant flavor constituents may be determined using this technique. Also, headspace sampling methods are not readily used with capillary
columns since the gas flow through these columns is quite low
(1-5 mL/min). The injection of a large volume of headspace
(1-2 mL) results in substantial band broadening of the early
eluting peaks and in very poor peak resolution of these very
volatile components. Therefore, most of the headspace analyses
on capillary columns was initially accomplished using the split
injection mode. However, there has been substantial interest
in applying larger headspace volumes to increase sensitivity
and yet maintain chromatographic resolution. To that end, this
study compares the sensitivity and reproducibility of split,
splitless and on-column injection techniques for the analysis
of headspace vapors.

The lack of sensitivity characteristic of split headspace injections is demonstrated in Table 2. None of the test compounds
were detected at 50 ppb and only the more volatile esters,
ethanol and 2-heptanone were detected at the 100 ppb level.
Since the detection limit of the GC is approximately 0.1 area
units and reproducibility typically is very poor at these low
levels, one would do best using this technique at the $\geq$ 1 ppm
level for even the very volatile constituents. Less volatile
components (e.g. ß-ionone and isoeugenol) do not even show up
at 10 ppm. This is well above their sensory thresholds.

Sensitivity could probably have been improved by making a slow injection and cold trapping in the column (1). However, there would still be the problem of sample loss through the split inlet system. Therefore, we chose to consider splitless head-space analysis.

An effort was made to determine the appropriate GC operating conditions and injection speed for splitless headspace. The role of injection speed in determining the recovery of vola-tiles is presented in Table 3. A fast injection results in very reduced GC response compared to slower injections. As Jennings and Rapp (1) pointed out, rapid injections of large headspace volumes results in the sample being lost to the in-let system. In the Hewlett Packard 5880, the carrier flow is directed through the septum purge line when operated in the splitless mode (prepurge time). Therefore, the fast injection forced a large portion of the sample into the septum purge system and directly out the vent line. An injection speed of 25 seconds/mL headspace was selected for the remainder of this project.

Table 2   Analysis of volatiles in aqueous systems using con-ventional Split (1:10) headspace techniques (5 repli-cates)

Compound	VOLATILE CONCENTRATION (60 min)								
	100 ppb			1 ppm			10 ppm		
	ng/mL[a]	Area[b]	CV[c]	ng/mL	Area	CV	ng/mL	Area	CV
Ethanol	0.3	.14	8.8	2.9	1.4	5.8	32	15	8.9
Propanol	*d	*	*	*	*	*	2.4	1.7	7.2
Butanol	*	*	*	*	*	*	0.7	2.4	15.4
Octane	*	*	*	0.7	.64	32.7	3.4	3.0	9.6
Decane	*	*	*	5.1	4.9	29.4	44	22	9.9
Ethyl Propanoate	0.9	.37	7.1	9.0	3.8	5.9	100	41	4.7
Ethyl Butanoate	*	*	*	21.9	5.4	4.4	238	59	2.6
Ethyl Pentanoate	1.3	.60	9.3	13.8	6.4	6.4	150	70	3.0

Cont. Table 2

2-Heptanone	0.6	.43	5.8	6.2	4.7	6.1	70	47	4.4
Acetophenone	*	*	*	*	*	*	3.2	2.5	21.1
Benzyl Acetate	*	*	*	2.2	0.87	13.4	24	10	18.5
Methyl Salicylate	*	*	*	*	*	*	2.2	3.9	42.9
L-Carvone	*	*	*	*	*	*	4.5	2.0	23.7
β-Ionone	*	*	*	*	*	*	*	*	*
Methyl Anthranilate	*	*	*	*	*	*	*	*	*
EtMePh Glycidate	*	*	*	*	*	*	*	*	*
Isoeugenol	*	*	*	*	*	*	*	*	*

[a]ng/ml headspace gas-average value; [b]Total peak area for 1 mL injection-average value, [c]Coefficient of variation; [d]Not detected

In practice, very good chromatography results with the use of splitless injection for liquid samples. The reason for good chromatography is that the solvent in the sample condenses at the head of the GC column and the sample recondenses in this solvent plug (10,11,12,13,14). What begins as a very slow vaporization and long injection time (adequate time for 1 1/2 inlet volumes to enter the column) becomes a rather sharp injection. In this study, we were curious whether this same phenomena could be reproduced with a headspace injection. Initially, we deciced to spike the sample bottle with from 5 to 25 μL of pure hexane. Then, when a headspace sample was drawn up into the syringe for analysis, hexane would also be included and it would act like the solvent of a typical liquid splitless injection. Unfortunately, the hexane did not adequately recondense the test volatiles and substantial band broadening resulted. As can be seen in Table 2, peak widths of the early eluting compounds were quite broad even when hexane was incorporated into the sample bottle. An alternative approach of using a dual injection, i.e. inject 0.5 μL liquid hexane and then immediately 1 mL headspace, did not signifi-

Table 3 Effect of various operating parameters on the recovery (GC Peak Area) of selected volatiles via the splitless injection method (1 ppm)

	SPLITLESS[a] Injection Speed				SPLITLESS[b] Injection				SPLIT[b] Fast Injection
	Fast 2 sec	5 sec	20 sec	Slow 30 sec	Fast[c] Hexane[d]	Fast[c] Dual[e]	Fast 2 sec	Slow 25 sec[f]	1:10
Ethyl Propano-ate	18	68	151	199	64	98(.04)g	96(.062)	255(.021)	13(.024)
Ethyl Butano-ate	29	111	235	324	93(.067)	133(0.058)	145(.074)	374(.025)	16(.037)
Ethyl Pentano-ate	*h	*	*	*	93(0.043)	145(.041)	171(.045)	442(0.24)	14(.032)
2-Heptanone	*	*	*	*	49(0.033)	86(.034)	82(.034)	240(.023)	7(.027)
Acetophenone	*	*	*	*	2(.018)	3(.018)	4	8(.018)	*

a Sample held at 60°C for 30 min.

b Sample held at 60°C for 60 min.

c 2 seconds to empty 1 mL headspace into GC inlet system.

d 20 µL of hexane was added to sample bottle prior to equilibration.

e 0.5 µL of hexane (liquid) was injected immediately prior to headspace injection.

f 20 cm of column was cooled with liquid nitrogen during injection.

g peak width at half height.

h Not determined.

cantly narrow the early eluting peaks. There may have been a minor improvement but not sufficient to warrant further study since the large hexane peak interferred with the quantitation of some of the test volatiles. The best means of narrowing peak was with cold trapping (Table 3). Combining a slow injection with cold trapping improved sensitivity of the headspace method by about 20-30 fold over the conventional split inlet headspace technique and significantly narrowed peak width.

Data demonstrating the sensitivity of splitless headspace sampling is presented in Table 4. One can see that detection limits of the more volatile test compounds are now in the ppt range. Even components as nonvolatile, as methyl salicylate (bpt. 220°C) and L-carvone (bpt. 230°C) can readily be detected at the 50 ppb level.

It is readily apparent that GC response does not proceed linearily with concentration. This is not unexpected since concentration in the headspace is related to solubility and mole fraction in the aqueous phase.

This method does show quite a high CV, especially at lower concentrations of test compounds. Errors in peak integration certainly increase as the total peak area decreases. There is also a problem with controlling syringe sample volume and injection rate. While sample loops are typically used in headspace sampling to improve precision, this technique does not lend itself to the use of a sample loop. The errors caused by imprecise rates of headspace injection are unique to this method (and on-column headspace). The use of internal standards would minimize most of the errors and substantially lower the CV.

The use of on-column headspace techniques offers many of the same advantages and disadvantages as does the splitless tech-

Table 4 Analysis of headspace vapors in aqueous system using splitless headspace techniques (4 replicates)

Compound	50 ppb			100 ppb			VOLATILE CONCENTRATION 1 ppm			10 ppm		
	ng/mL[a]	Area[b]	CV[c]	ng/mL	Area	CV	ng/mL	Area	CV	ng/mL	Area	CV
Ethanol	9.0	38	72	15	64	52	37	158	6.4	407	1723	25
Propanol	1.2	6.5	13	2.0	9	67	5.8	26	19	50	224	25
Butanol	0.6	3.3	173	1.2	6.3	112	6.0	31	20	61	324	24
Octane	17	109	56	22	146	80	26	168	62	170	1103	22
Decane	42	443	48	43	440	30	55	565	75	760	7765	53
Ethyl Propanoate	4.3	32	15	7.5	56	18	64	481	19	580	4321	18
Ethyl Butanoate	7.8	68	18	12	105	26	80	699	15	730	6384	14
Ethyl Pentanoate	20	134	24	25	172	32	130	821	13	1100	7642	9
2-Heptanone	8.0	82	22	9.8	95	36	53	512	20	490	4700	20
Acetophenone	1.0	14	16	0.7	10	66	2.0	28	30	14	185	26
Benzyl Acetate	2.9	29	57	3.9	39	44	7.6	76	22	56	559	18
Methyl Salicylate	1.5	12	61	1.8	15	23	3.7	31	25	24	203	6.9
L-Carvone	0.6	8	43	0.8	11	22	2.1	29	16	13	175	11
ß-Ionone	1.8	19	104	2.4	26	74	7.0	74	16	62	662	30
Methyl Anthranilate	0.02	.60	69	0.09	1.1	61	0.03	20	73	3.1	36	34
EtMePh Glycidate	*[d]	*	*	*	*	*	0.6	3.8	120	2.1	13	50
Isoeugenol	*	*	*	*	*	*	0.6	5.8	79	1.8	18	24

[a] ng/mL headspace; [b] Average GC peak area; [c] Coefficient of variation; [d] Not determined

Table 5   Analysis of volatiles in aqueous model systems using on-column
Headspace techniques (5 replicates)

Compound	VOLATILE CONCENTRATION								
	100 ppb			1 ppm			10 ppm		
	ng/mL[a]	Area[b]	CV[c]	ng/mL	Area	CV	ng/mL	Area	CV
Ethanol	*	*	*	2.6	97	13.4	4.8	181	9.8
Propanol	0.4	1.0	25	2.0	4.8	11.5	19	46	70.0
Butanol	*	*	*	1.2	3.2	17.9	17	46	27.9
Octane	0.7	.9	32	3.0	38	21.2	11	141	38.1
Decane	0.6	2.0	43	37	126	7.6	3.6	124	28.4
Ethyl Propanoate	2.8	5.0	13	36	62	12.6	488	849	13.5
Ethyl Butanoate	4.6	8.5	7.6	48	87	6.0	607	1101	16.9
Ethyl Pentanoate	6.0	12	4.5	47	93	7.2	550	1087	20.0
2-Heptanone	2.2	6	4.6	26	71	6.9	308	845	21.8
Acetophenone	*	*	*	0.8	2.8	12.0	11	37	24.8
Benzyl Acetate	0.6	1.6	8.2	4.4	12	15.5	44	117	41.3
Methyl Salicylate	0.4	0.9	13	1.8	4.2	16.4	18	42	25.9
L-Carvone	*	*	*	0.8	3.0	12.1	16	28	22.1
β-Ionone	*	*	*	*	*	*	22	47	20.5
Methyl Anthranilate	*	*	*	*	*	*	1.4	2.8	20.9
EtMepH Glycidate	*	*	*	*	*	*	1.6	1.2	25.1
Isoeugenol	*	*	*	*	*	*	1.0	1.6	57.2

[a] ng/mL headspace; [b] GC peak area/mL headspace; [c] Coefficient of variation; [d] Not detected

nique. However, the on-column method does insure that the sample contacts the most inert environment and is vaporized under the mildest conditions possible. On-column headspace systems and syringes have been developed which contain no metal/sample contact. While thermal instability and reactivity were not considerations in this study, some common flavor constituents are thermally labile (terpenes) and reactive (phenols and sulfur compounds). Therefore, the use of on-column headspace techniques may well be justified in many flavor studies.

A demonstration of the precision of on-column headspace sampling is presented in Table 5. The method showed quite low CV for the different concentrations examined.

A comparison of split, splitless and on-column headspace sampling methods is presented in Table 6.

It is obvious that both splitless and on-column techniques were substantially more sensitive than the conventional split headspace method. Under the conditions used in this study, the splitless method was lightly more sensitive than the on-column method. There is no inherent reason for this. Simply more time was spent optimizing the GC system and injection technique for the splitless method. We would expect the two methods to yield comparable results when both are optimized.

From an operator standpoint, the on-column technique does require more skill to use. Certain seals and valves must be maintained in good working order. However, the additional effort may often be offset by the "gentleness" of the on-column technique.

Table 6   Comparison of peak areas observed for split, splitless and on-column methods of headspace analysis
(1 ppm)[a]

| | INJECTION METHOD | | |
Compound	Split 1:10 Fast[b]	Splitless Cold Trapped Slow	On-colum Cold Trapped Slow
Butanol	*c	31	18
Decane	*	565	401
Ethyl Propanoate	24	481	390
Ethyl Butanoate	46	699	492
Ethyl Pentanoate	36	821	610
2-Heptanone	17	512	392
Acetophenone	2	28	18

[a]Sample equilibrated at 60°C for 1 hr.
[b]Rate of injection.
[c]Not detected.

Freon Reflux

The use of Freon 12 for headspace vapor extraction was originally proposed by Jennings (6). He demonstrated the potential
of this technique by presenting some work on the volatiles of
banana and a chili pepper spice preparation. While the apparatus used in this study is basically the same as used by Jennings
(6), we decided to add 2 mL of diethyl ether containing internal standard at the end of the trapping period but prior to
Freon evaporation. This permitted efficient transfer of the
trapped volatiles into a vial for storage until analysis, left
a solvent for the volatiles and provided a convenient means of
adding internal standard for quantitation. The addition of diethyl ether also provided a convenient means of adding inter-

nal standard for quantitation. The addition of diethyl ether
also provided a simple means of obtaining a moisture free sample. Since Freon 12 boils at -30°C, the sample collection flask
remained cold until the Freon had completely evaporated. Therefore, the ether was pipetted out of the sample flask when the
Freon had evaporated but before it warmed sufficiently to melt
the ice. The result was a moisture-free sample for GC analysis.

The recovery of test compounds from aqueous solution (50 ppb)
using Freon extraction is shown in Table 7. An examination of
the table shows that quantitative relationships between purge
times are quite variable. One would expect the 3 hr purge to
contain 3 X the volatiles as the 1 hr purge. This is assuming
that the concentration of volatiles in the purge gas is uniform with time.

Table 7   Recovery of volatiles from an aqueous system (50 ppb)
          using headspace concentration via Freon reflux (2 re-
          plicates)

	PERCENT RECOVERY OF VOLATILES			
	--------21°C-----Purge Temperature------- 60°C-------			
Compound	1 hr[a]	3 hr	1 hr	3 hr
Ethanol	*[b]	*	.64	.26
Propanol	*	.12	1.21	.55
Butanol	*	*	.23	.93
Octane	*	*	*	.85
Decane	*	.15	*	.17
Ethyl Propanoate	.27	1.85	2.75	3.73
Ethyl Butanoate	.45	2.84	4.03	4.80
Ethyl Pentanoate	.79	3.25	4.25	4.91
2-Heptanone	.19	1.44	2.52	3.81
Acetophenone	*	.15	.26	.52
Benzyl Acetate	.07	.64	1.14	3.19
Methyl Salicylate	*	.36	.48	1.22

Cont. Table 7

L-Carvone	*	.11	.22	1.18
ß-Ionone	*	.17	*	.69
Methyl Anthrani-late	*	*	*	*
EtMePh Glycidate	*	*	*	*
Isoeugenol	*	*	*	*

[a]Purge time
[b]Not detected

However, our data indicate there was less ethanol and propanol and only slightly more of the esters recovered after 3 hrs of purging compared to the 1 hr purge. This suggests that we had problems with control of the extraction and perhaps trapping efficiency in the apparatus. Qualitatively, Freon reflux was substantially better than split headspace and comparable to splitless and on-column headspace.

The primary advantage of this technique is that one can obtain a headspace sample which is moisture free and virtually sol-vent free (if ether is not added). Additional advantages are that sufficient sample is obtained to make multiple injections onto the GC or mass spectrometer and the low boiling solvent is ideal for the recovery of very volatile flavor components. The primary disadvantage of the method is the problem of work-ing with a solvent which boils below room temperature. This makes distillation (to purify), storage and handling of the solvent as well as control of the distillation (or reflux) more difficult.

Adsorbent Trapping

Volatile flavors are often trapped out of purge gas via adsorbent traps since these adsorbents typically have little affinity for water but a high affinity for organic molecules. Therefore, a flavor isolate is obtained which is free of solvent and water. We chose to evaluate the use of Tenax as the adsorbent material. The recoveries obtained using a commercial purge and trap system are presented in Table 8. Excellent recoveries and very low CV were observed for the more volatile esters. As expected, recovery of the higher boiling compounds was substantially poorer.

While purge times of 15 and 30 min would have been sufficient, it was of interest to look for "break through" of some of the components. A comparison of the 30 min and 1 hr purge times suggests that the alcohols and esters did break through the trap. The quantities of these compounds recovered was substantially less than expected if no break through had occurred.

Overall, the purge and trap data indicate that this is a very sensitive method capable of detecting some compounds in the ppt range. For example, ethyl butanoate gave a GC peak area of over 4,000 at 50 ppb. There is ample sensitivity to detect ppt levels of this compound. We should point out that all purging was done at room temperature. We expect that 60°C purging would have resulted in even better sensitivity and perhaps the detection of some of the higher boiling compounds. The CV on this method was also quite low for most compounds. The total automation and closed system approach is responsible for this excellent precision.

Solvent Extraction

Solvent extraction offers a very simple and efficient means of isolating flavors from foods. Unfortunately, this technique is limited to the analysis of foods which contain no lipids unless a secondary isolation method (e.g. dialysis or distillation) is

Table 8 Recovery of volatiles from an aqueous model system (50 ppb) using adsorption traps and a purge and trap sampling system (5 replicates)

Compound	15 min purge			30 min purge			1 hr purge[a]	
	μg[b]	Recovered %	CV	μg	Recovered %	CV	μg	Recovered %
Ethanol	1.6	3.2	4.3	1.6	31.1	21.7	.79	1.5
Propanol	.63	1.2	11.1	1.2	2.5	6.4	.75	1.5
Butanol	.58	1.3	15.9	1.6	3.5	9.4	3.1	6.7
Octane	*	*	*	*	*	*	*	*
Decane	*	*	*	*	*	*	*	*
Ethyl Propanoate	20.8	44.3	20.4	43.1	91.9	1.7	49.8	105.9
Ethyl Butanoate	24.5	51.0	23.1	50.0	104.0	1.2	52.0	108.3
Ethyl Pentanoate	12.9	38.8	27.6	38.0	82.6	0.6	36.8	80.0
2-Heptanone	11.3	26.9	11.3	20.8	49.4	2.8	30.3	72.2
Acetophenone	1.1	2.1	9.1	2.5	4.7	3.0	6.8	12.8
Benzyl Acetate	2.6	4.6	37.3	14.4	25.7	8.4	20.3	35.8
Methyl Salicylate	1.8	2.7	17.5	9.5	14.4	8.8	28.4	58.2
L-Carvone	1.0	1.9	225	2.9	5.7	4.8	7.5	14.4
β-Ionone	*	*	*	1.1	0.2	29.6	0.4	*
Methyl Anthranilate	*	*	*	*	*	*	*	*
EtMePh Glycidate	*	*	*	*	*	*	*	*
Isoeugenol	*	*	*	*	*	*	*	*

[a]Single determination; [b]μg recovered by sampling system (value x 100); [c]Not detected

also applied. This study evaluates the efficiency of pentane,
diethyl ether and dichloromethane as extracting solvents when
used in both batch and continuous processes.

The batch method is very simple but labor intensive. Vigorous
shaking for considerable time is required to make the method
effective. However, as can be seen in Table 9, very high recoveries and low CV can be obtained via this method. In this
study, dichloromethane was found to be the best solvent followed by pentane and then ether. Extraction efficiencies were
substantially higher for dichloromethane as compared to either
ether or pentane extractions. This is in agreement with the
work on model systems by Cobb and Bursey (15).

There are some noticeable differences in the proportions of
compounds extracted by each solvent. Pentane is a very poor
solvent for the extraction of polar substances (e.g. alcohols).
This may be used to advantage when extracting flavors from
aqueous/alcohol mixtures (16,17,18,19).

Two disadvantages of solvent extraction apparent from the data
on low boilers. For the dichloromethane extraction, four of the
test compounds were lost in the solvent front. Only two compounds were lost with the ether and one with the pentane. Therefore, the low boilers may be lost in the solvent front when a
solvent extraction technique is used for flavor isolation. A
second problem with low boilers is that they may be lost during
concentration of the solvent. While recovery would have been
improved we had used a Kuderna-Danish concentrator for solvent
removal or a spinning band distillation system, it is reasonable
to expect some losses of low boiling volatiles during the concentration step. These losses may be minimized via the use of
very low boiling solvents (e.g. liquid CO_2 or fluorocarbons)
and more efficient concentration equipment.

Continuous extraction methods may be used to replace the more

Table 9 Recovery of volatiles from an aqueous model system (50 ppb) via solvent extraction (5 replicates)

	BATCH EXTRACTION						CONTINUOUS EXTRACTION							
	Dichloromethane %[a]		Ether %		Pentane %		Dichloromethane % 2 hr		Dichloromethane % 4 hr		Dichloromethane % 2 hr		Diethyl Ether % 4 hr	
	Recovery	CV	Recovery	CV	Recovery	CV	Recovery	CV	Recovery	CV	Recovery	CV	Recovery	CV
Ethanol	*[b]	*	-	-	*	*	*	*	*	*	45.9	44.4	46.8	25.8
Propanol	*	*	6.32	6.88	*	*	*	*	*	*	9.86	17.1	16.7	25.6
Butanol	28.9	7.84	27.4	4.03	-	-	22.5	5.31	63.6	15.7	40.0	10.5	52.3	15.0
Octane	-	*	-	-	0.54	11.4	*	*	*	*	-	-	-	-
Decane	-	-	-	-	7.17	13.7	-	-	-	-	-	-	-	-
Ethyl Propanoate	*	*	24.6	2.48	17.7	7.94	*	*	*	*	36.3	12.1	38.9	7.79
Ethyl Butanoate	62.7	8.00	28.9	5.73	41.6	10.2	43.1	2.29	69.7	24.0	41.8	10.0	46.2	12.4
Ethyl Pentanoate	64.6	9.24	32.5	6.05	53.4	4.58	44.6	2.18	57.1	21.3	47.3	11.8	47.9	7.99
2-Heptanone	74.4	7.80	39.3	1.92	53.8	5.79	48.6	2.39	59.8	6.3	53.8	10.6	54.5	5.52
Acetophenone	100.1	6.73	53.4	3.85	65.9	6.95	71.6	5.75	82.3	10.5	74.8	11.9	73.6	4.70
Benzyl Acetate	81.4	7.44	45.8	3.09	64.6	4.78	51.3	4.80	76.3	5.2	63.8	8.46	62.1	4.48
Methyl Salicylate	84.7	6.87	42.1	3.85	67.3	4.61	54.0	5.09	81.8	4.5	60.0	9.51	56.9	4.80
L-Carvone	94.9	7.91	46.8	5.33	60.0	9.25	59.3	5.43	99.0	6.7	64.2	14.1	68.3	10.9
β-Ionone	77.5	8.09	21.6	9.58	39.0	16.0	46.8	5.10	84.5	10.9	34.8	22.8	26.6	16.5
Methyl Anthranilate	90.9	7.48	36.5	11.3	7.0	28.8	65.2	11.2	101.2	12.8	38.9	14.9	33.3	14.1
EtMePh Glycidate	99.2	9.00	35.0	7.89	18.9	29.5	58.2	6.28	112	9.8	45.0	12.3	40.1	12.8
Isoeugenol	77.6	26.9	2.36	2.36	41.1	18.3	66.3	13.1	91.9	50.7	14.5	44.4	13.1	75.2

[a] Mean value
[b] Lost with solvent front

tedious batch extraction method. The continuous extractors
typically are reflux systems where the solvent is distilled
and then flows either up (lighter-than-water) or down (heavier-
than-water) through the sample. The solvent plus extractables
return to the distillation pot for continued extractions.

A question often arises concerning the length of time the continuous extractor should be operated. In this study we chose
to evaluate 2 and 4 hr extraction times. The ether and dichloromethane yielded very similar results for the 2 hr extractions,
averaging approximately 50% recoveries. The 4 hr data on dichloromethane extraction shows quite good recoveries. Note that
the recovery of low boilers was improved compared to the batch
process due to the lower solvent volume used in the continuous
extractor. The 4 hr data on ether extraction seems unreasonable.
We cannot account for the similarity in 2 hr and 4 hr data.
However, these results were reproducible.

In summary, dichloromethane appears to be the solvent of choice
in this study followed by pentane and diethyl ether. Batch extraction was found to yield comparable recoveries of test compounds to continuous extraction. It is estimated that 6 hr
should be adequate to extract most volatiles from foods via
continuous extraction. While the batch process was tedious, it
did require less total time for extraction than did continuous
extraction (30 mins vs. 5 or 6 hr).

Simultaneous Distillation/Solvent Extraction

Simultaneous distillation/solvent extraction (SDE) offers a
convenient means of obtaining flavor isolates from foods containing lipids and therefore, cannot be directly solvent extracted. The SDE head used in this work is that designed by
Schultz et al. (7) but modified to include a cold trap (dry
ice/ethanol) at the vacuum connection of the apparatus. This

additional cold trap was required in order to maintain the
solvent in the apparatus when vacuum was applied.

Recoveries of test compounds via the SDE method are presented
in Table 10. Compounds which chromatographed from butanol to
L-carvone exhibited the best recoveries. The low boiling volatiles would be expected to be recovered in higher proportions
if not for losses during solvent concentration. The high boilers would not be recovered well due to their low vapor pressure.

Recovery of many of the test compounds changed very little as
a function of extraction time. It would appear that an equilibrium may have been established which prohibited further extraction. Only some of the very late eluting volatiles continued to improve in recovery as longer extraction times were
used (e.g. ß-ionone, methyl anthranilate, EtMePh glycidate and
isoeugenol).

Going to a vacuum SDE had a very small effect on recovery. Overall there appeared to be a slight negative effect on recovery.
While vacuum operation is not needed in this study, vacuum
operation is required on real food systems to minimize artifact formation due to thermally induced changed.

Summary

A summary of selected data is presented in Fig. 1. This graphical presentation illustrates the differences in sensitivity
between the methods used in this study.

The dependency of the headspace methods on volatility is obvious.
Recovery drops off quite quickly as vapor pressure of the compound in aqueous systems decreases. It is evident that both
splitless and on-column headspace methods offer vast improvements in sensitivity over the conventional split injection head-

Table 10 Recovery of volatiles from an aqueous model system (50 ppb) using simultaneous distillation/solvent extraction (5 replicates)

| | ATMOSPHERIC | | | | | | REDUCED PRESSURE 400 mm Hg | | | |
| | 1 hr | | 2 hr | | 4 hr | | 2 hr | | 4 hr | |
	Mean % Recovery	CV	Mean % Recovery	CV	Mean % Recovery	CV	Mean % Recovery	CV	Mean % Recovery	CV
Ethanol	−	−	−	−	−	−	−	−	−	−
Propanol	19.6	9.89	28.5	14.5	36.8	11.7	32.0	9.32	31.2	5.95
Butanol	43.6	11.4	59.9	16.6	61.6	5.79	51.7	7.80	54.9	10.3
Octane	0.91	27.1	0.87	63.4	−	−	12.0	63.2	23.4	35.3
Decane	−	−	−	−	−	−	−	−	−	−
Ethyl Propanoate	39.1	3.39	48.0	7.49	43.7	5.21	38.6	8.14	38.9	5.73
Ethyl Butanoate	41.1	31.8	49.7	9.02	46.8	12.7	40.0	8.40	41.2	4.99
Ethyl Pentanoate	43.1	5.79	51.3	10.2	48.3	6.68	42.3	7.68	44.6	6.69
2-Heptanone	53.2	6.58	65.5	7.79	57.7	4.42	49.2	7.33	50.3	6.79
Acetophenone	58.3	9.20	76.1	3.49	71.4	3.50	57.6	8.03	57.2	8.79
Benzyl Acetate	52.5	9.20	61.4	3.38	59.2	3.17	49.8	8.53	49.7	8.47
Methyl Salicylate	42.6	14.9	46.7	5.22	47.8	6.89	47.0	9.17	47.2	9.56
L-Carvone	54.5	9.80	59.6	6.36	66.2	3.72	56.4	9.32	54.1	11.7
β-Ionone	30.6	15.4	28.8	21.1	41.8	13.8	49.4	23.39	34.0	26.7
Methyl Anthranilate	10.7	19.4	16.1	23.4	34.8	9.82	21.8	27.70	27.3	22.2
EtMePh Glycidate	16.6	18.6	17.2	28.2	27.9	16.1	26.6	23.59	29.2	26.8
Isoeugenol	4.19	30.0	3.87	25.9	5.66	38.4	3.10	58.50	18.9	93.7

45

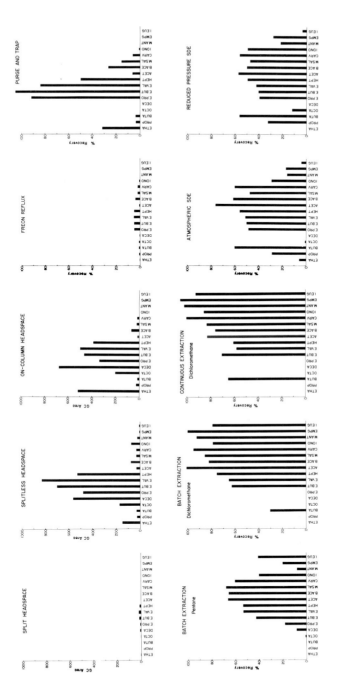

Fig. 1 Comparison of selected methods for the recovery of volatiles
from aqueous systems (headspace methods 1 ppm - all others 50 ppb)

46

space method. Some of the volatiles studied would have detection limits in the ppt range using either of these more sensitive techniques.

The Freon reflux and adsorbent trapping method yielded qualitatively similar data to the splitless and on-column headspace methods. However, the adorbent method did extend sensitivity to even lower limits and exhibited very low CV. The Freon reflux technique was of particular interest since it yielded an isolate potentially solvent free and yet was adequate in quantity to permit repeated injections. This technique is the only "headspace" method that did recover octane and decane.

The most efficient method for flavor isolation was solvent extraction. Even at 50 ppb, excellent recoveries were observed. The problem of the solvent front overlapping with low boilers is particularly obvious with the dichloromethane solvent. While dichloromethane yielded the best recoveries, it also prohibited quantitation of several of the low boilers. The effect of solvent polarity on recovery is very noticeable for the pentane extraction. The polar alcohols and somewhat water soluble high boilers were poorly recovered.

Simultaneous distillation/solvent extraction (SDE) methods exhibited very good recoveries. Since foods which contain lipids cannot be directly solvent extracted, SDE methods offer an efficient alternative.

References

1. Jennings, W.G., Rapp, A.: Sample Preparation for Gas Chromatographic Analysis, Hüthig Publ., New York 1983.
2. Reineccius, G.A., Anandaraman, A.: Recent Advances in the Chromatographic Analysis of Organic Compounds in Foods, Marcel Dekker Inc, New York 1983.

3. Maarse, H., Belz, R.: Handbuch der Aromaforschung, Akademie-Verlag, Berlin 1981.

4. Teranishi, R., Flath, R.A., Sugisawa, H.: Flavor Research,
   Recent Developments, Marcel Dekker Inc., New York 1981.

5. Bemelmans, J.M.H., in: Progress in Flavour Research (Land,
   D.G., Nursten, H.E., Eds.), Applied Science Publ., London
   1979.

6. Jennings, W.G.: HRGC & CC 2, 221 (1979).

7. Schultz, T.H., Flath, R.A., Mon, T.R., Eggling, S.B., Teranishi, R.: J. Agric. Food Chem. 25, 446 (1977).

8. Jennings, W.G., Filsoof, M.: J. Agric. Food Chem. 25, 440
   (1977).

9. Schaefer, J., in: Handbuch der Aromaforschung (Maarse, H.,
   Belz, R., Eds.), Akademie-Verlag, Berlin 1981.

10. Jennings, W.G.: Gas Chromatography with Glass Capillary
    Columns, Academic Press, New York 1980.

11. Grob, K., Grob, G.: Chromatographia 4, 3 (1972).

12. Grob, K., Grob, K. jr.: J. Chromatogr. 94, 53 (1974).

13. Grob, K., Grob, K. jr.: HRC & CC 1, 57 (1978).

14. Jennings, W.G., Freeman, R.R., Rooney, T.R.: HRC & CC 1,
    275 (1978).

15. Cobb, C.S., Bursey, M.M.: J. Agric. Food Chem. 26, 197
    (1978).

16. Schreier, P.: J. Agric. Food Chem. 28, 926 (1980).

17. Schreier, P., Drawert, F., Winkler, F.: J. Agric. Food Chem.
    27, 365 (1979).

18. Tressl, R., Friese, L., Fendesack, F., Köppler, H.: J.
    Agric. Food Chem. 26, 1422 (1978).

19. Tressl, R., Friese, L., Fendesack, F., Köppler, H.: J.
    Agric. Food Chem. 26, 1426 (1978).

SELECTIVE ENRICHMENT OF VOLATILES BY GAS-WATER PARTITION IN
CON- AND COUNTERCURRENT COLUMNS

Otto Piringer and Heinrich Sköries
Fraunhofer-Institut für Lebensmitteltechnologie und Verpackung
Schragenhofstr. 35, D-8000 München 50

Introduction

A general characteristic of analysis at low concentration
ranges is the considerable variety of compounds. Therefore,
even with high resolution chromatographic separations peak
overlapping may appear, and selective sample enrichment is
often the sole chance for problem solution. Sample partition
between aqueous solutions and a gas offers a tool for this aim
as the partition coefficients cover a large interval of values.
Whereas head space and various gas stripping techniques (1-4)
with limited liquid volumes are frequently used in analysis,
columns working under steady state conditions are less widespread despite of their use at technical scales. A supposition
of enlarged applications of partition processes in selective
enrichments is a data collection of gas-water partition coefficients. An explanation of the present lack of data lies in
the many sources of errors in their measurements at high dilutions, especially errors caused by adsorption (5-13).

A dynamic method and a column are now described, which allow
both the determination of partition coefficients in air-water
systems and selective enrichment of organics by gas stripping
from aqueous solutions. With the operation mode under steady
state conditions using practically unlimited liquid and gas
volumes, errors from adsorption processes can be largely eliminated. The height equivalent of a theoretical plate is only

a fraction of the effective column length, so phase equilibrium is guaranteed at the column exit in the concurrent mode.

Theory

The partition coefficient $K_{g/l}$ of a solute between a gas (g) and a liquid(l)is defined as the ratio of the equilibrium solute concentration in the gas phase C_g and in the liquid phase C_l. It is a function of the vapour pressure for a pure compound p^o and the activity coefficient of the substance in the liquid phase γ_l expressed on a Raoult's law condition:

$$K_{g/l} = \frac{C_g}{C_l} = \gamma_l \cdot p^o \cdot \frac{V_l}{RT} \qquad (1)$$

V_l, T and R are the molar volume of the liquid phase, the temperature and the gas constant, respectively. For aqueous solutions, l = w, and 25°C the equation (1) becomes:

$$K_{g/w} = 9.681 \cdot 10^{-7} \cdot \gamma_w \cdot p^o \qquad (2)$$

with p^o in mm Hg.
In order to obtain $K_{g/w}$ the vapour pressure can be calculated from the Antoine equation:

$$\log p^o = A - \frac{B}{t + C} \qquad (3)$$

where A,B,C are correlation constants and t the temperature (°C). On the other hand γ_w can be calculated if the solute solubility in water $C_w^{(m)}$ (mol/l) is known:

$$\gamma_w = \frac{1}{C_w^{(m)} \cdot V_w} \qquad \text{for liquids}$$

$$\gamma_w = \frac{1}{C_w^{(m)} \cdot V_w} \cdot \exp\,[6,79\,(1 - Tm/T)] \qquad (4)$$

for solids

If the solute is solid at the investigated temperature T (K)
a correction has been introduced (14) with the solid's melting
point T_m (K). A further possibility of calculus for γ_w results
from the near inverse relationship between solubility and the
octanol-water partition coefficient $K_{o/w}$ (14):

$$\ln\,C_w^{(m)} = 7.494 - \ln\,K_{o/w} \qquad \text{for liquids}$$

$$(5)$$

$$\ln\,C_w^{(m)} = 7.494 - \ln\,K_{o/w} + 6.79\,(1 - Tm/T)$$

for solids

Again, a correction for solid solutes has improved the correlation. The equations above open possibilities for data correlations from different experimental sources, for a check of
systematic errors and for data prediction.

Experimental

Figure 1 shows the essential part of a column (C) for direct
measurements of partition coefficients. The liquid phase with
an initial solute concentration $C_{1,o}$ runs at the flow rate L
(0.1-0.3 l/h) along a vertical glass helix (S) mounted around
the inner tube of two concentric glass tubes. A stream of air
or nitrogen flows at the rate G (10-40 l/h) in concurrent with
the liquid through the phase contactor room (D) between the
two concentric tubes. Thermostated water flows around the outer
tube and through the inner tube and holds the column at constant temperature T. The necessary contact surface and time

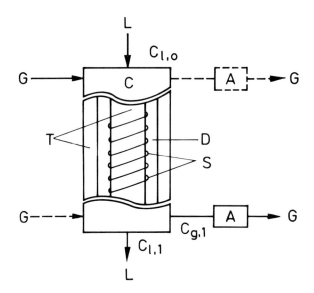

Fig. 1   Column for gas-liquid desorption in concurrent (——)
and countercurrent (- - -)

for phase equilibration are established with the $C_{g,1}$ the gas
phase passes through cooled pentane or another suitable solvent (A) where the solutes are absorbed. The liquid phase
leaves the column with the final solute concentration $C_{l,1}$.
The partition coefficient is obtained from the measured ratio
$C_{g,1}/C_{l,1}$. If complex solutions of unknown composition are investigated and solute analysis in the liquid phase is not possible without difficulties, two successive runs with the same
liquid permit the determination of the partition coefficient
from $C_{g,1}$ in the first and $C_{g,2}$ in the second run:

$$K_{g/w} = \frac{L_2}{G_2} \left(\frac{C_{g,1}}{C_{g,2}} - 1 \right) \tag{6}$$

L_2/G_2 is the phase flow ratio in the second run.

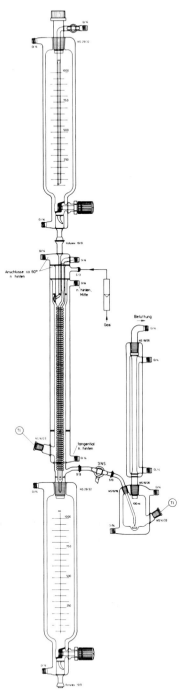

Fig. 2 Apparatus for the determination of gas-liquid partition coefficients and selective enrichment

Fig. 2 shows a more detailed picture of the whole apparatus.
(It is offered by GREINER & GASSNER GmbH Glastechnik, Kathi-
Kobusstr. 15, D-8000 München 40 and OTTO FRITZ GmbH Normschliff-
Aufbaugeräte, Feldstr. 1, D-6238 Hofheim im Taunus).
In order to concentrate the pentane solution the absorption
vessel can be connected with a spinning band distillation
column (Normag).

Table 1   Air-water partition coefficients of some volatiles
at 25°C

Compound	$K_{g/w} \cdot 10^3$		
	this work	other authors	
Hexanol	0.7 ± 0.1	0.7 ± 0.06	(6)
Heptanol	0.9 ± 0.1	-	
Pentan-2-one	3.2 ± 0.4	2.6 ± 0.2	(6)
Octan-3-one	8.8 ± 0.6	-	
Pentan-2,4-dione	2.0 ± 0.2	-	
Oct-trans-2-enal	4.0 ± 0.4	3.0	(7)
Methyl pentanoate	12.5 ± 0.8	13 ± 3	(6)
Pentyl acetate	15 ± 1	14.5 ± 0.2	(9)
tert-Butyl propionate	54 ± 5	-	

Table 2   Solubilities and air-water partition coefficients of
some PCB's measured at 25°C

PCB	C_w mg/kg	$\gamma_w \cdot 10^{-7}$	$K_{g/w} \cdot 10^2$	$p^o \cdot 10^4$ mm Hg (calc)
2,2' -dichloro-	1.1	1.1	1.0	0.9
2,3 -dichloro-	0.44	2.8	1.6	0.6
4,4' -dichloro-	0.32	3.9	1.4	0.4

Cont. Table 2

2,2',3-trichloro-

                  0.17   8.4   3.5   0.4

2,2',4-trichloro-

2,3,3'-trichloro-
2,3,4'-trichloro- 0.07   20   1.8   0.1
2,4,4'-trichloro-

Results and Discussion

In Table 1 some air-water partition coefficients measured
with this column at 25°C are given.
Interposing a column containing the investigated substance
deposited on fine glass beads (15) between the liquid phase
reservoir and the desorption column (Fig. 2), both the solubility and the air-water partition of the substance could be
measured. Table 2 shows the results for some dichloro- and
trichloro-biphenyls. With equation 4 the activity coefficient
γ_w can be calculated from the solubility and with equation 1
the low vapour pressures of these compounds can be estimated.

The activity coefficient γ_w is a measure of hydrophobicity and
approximately its logarithm increases linearly with the molecular size. As a consequence the partition coefficient $K_{g/w}$
increases in a homologous series with solute's molecular weight
M as already PIEROTTI et al. (16) have shown in 1959, whereas
in less polar solutions a decrease is registered (Fig. 3).
This fact, together with the definition of a gas chromatographic separation factor $\alpha_{j,i}$ for two solutes i and j:

$$\alpha_{j,i} = \frac{t'_R (j)}{t'_R (i)} = \frac{K_{1/g} (j)}{K_{1/g} (i)} = \frac{\gamma_1 (i) \cdot p^o_i}{\gamma_1 (j) \cdot p^o_j} = \frac{K_{g/1} (i)}{K_{g/1} (j)} \quad (7)$$

and the molecular retention index Me (17) for a solute i on the liquid stationary phase p:

$$Me_i^{(p)} = 14 \frac{\log K_{1/g} (i) - \log K_{1/g} (n)}{\log K_{1/g} (n+1) - \log K_{1/g} (n)} + M_n = 0.14 I_i^{(p)} + 2$$

$$(8)$$

a corresponding molecular retention index for water as solvent is defined:

$$Me_i^{(w)} = -14 \frac{\log K_{g/w} (i) - \log K_{g/w} (n)}{\log K_{g/w} (n+1) - \log K_{g/w} (n)} - M_n \quad (9)$$

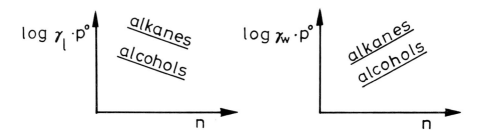

Fig. 3   The partition coefficients in homologous series for organic solvents (1) and water (w)

This definition takes into account the increase of $K_{g/w}$ within a homologous series mentioned above. Let us express the quantity $Me_i^{(p)}$ as sum of the relative molecular weight M_i and a relative retention weight $W_i^{(p)}$ accounting for the solute's polarity in the liquid stationary phase p (18). Let us further express for aqueous solutions, the quantity $Me_i^{(w)}$ as difference of a structure increment $W_i^{(w)}$ and M_i. Then an useful connection between the adjusted retention time t'_R and KOVATS index $I_i^{(p)}$ on the phase p (equation 7 and 8) and the air-water partition coefficient $K_{g/w,i}$ (equation 9) can be established:

$$Me_i^{(C-20\ M)} = W_i^{(C-20\ M)} + M_i$$

$$\tag{10}$$

$$Me_i^{(w)} = W_i^{(w)} - M_i$$

For Carbowax 20-M as an example many data of KOVATS indices and $W_i^{(C-20M)}$ are available (18,19) and in Table 3 the corresponding $W_i^{(w)}$ values are given for few classes of compounds. $I_i^{(C-20M)}$ and $K_{g/w,i}$ and consequently $Me_i^{(C-20M)} + Me_i^{(w)} = W_i^{(C-20M)} + W_i^{(w)}$ allow prediction of the nature of solute i and an estimation of its molecular weight M_i.

Table 3   Some polarity increments for water and Carbowax-20M

Class of compounds	$W^{(w)}$	$W^{(C-20M)}$
n-Alkanes	0	0
1-Alkenes	74	8
1-Chloroalkanes	190	30
Monoalkyl benzenes	231	60
Aldehydes	359	54
Esters	364	35 acetates
Ketones	390	53
p-Alcohols	466	85

The tabulated values of $W^{(w)}$ were calculated on the basis of the air-water partition coefficients of n-hexane, n-heptane and n-octane, very carefully measured recently (13). They allow prediction of the partition coefficients with the formula:

$$\log K_{g/w \ (i)} = \log 101 - \frac{0.157}{14} \ (W_i^{(w)} - M_i + 100) \tag{11}$$

where log 101 comes from the partition coefficient of n-heptane and 0.157 from the logarithm of $K_{g/w \ (n+1)}/K_{g/w \ (n)}$. (n = carbon number of alkane)

A further advantage of expressing these quantities on a molecular weight base is their connection to mass spectrometry where the ion molecules and all fragments for structure identification are expressed in relative molecular weight units.

The values of air-water partition coefficients are spread out between limits which differ by several orders of magnitude and this is the base for selective enrichment by gas stripping of aqueous solutions. The ratio Q of the desorbed amount of a solute to the initial amount in the liquid phase can be calculated for one equilibration stage with the formula:

$$Q = \frac{G \cdot K_{g/l}}{G \cdot K_{g/l} + L} \tag{12}$$

and is a function of the flow ratio G/L and $K_{g/w}$ (Table 4).

Table 4   Percentage of desorption in dependence on K and G/L in one equilibration stage

K	G/L	$100 \cdot Q$	K	G/L	$100 \cdot Q$
$1 \cdot 10^{-5}$	100	0.1	$1 \cdot 10^{-5}$	10^4	9
$1 \cdot 10^{-4}$	100	1	$1 \cdot 10^{-4}$	10^4	50
$1 \cdot 10^{-3}$	100	9	$1 \cdot 10^{-3}$	10^4	91

Cont. Table 4

$1 \cdot 10^{-2}$	100	50		$1 \cdot 10^{-2}$	10^4	99
$1 \cdot 10^{-1}$	100	91				
1	100	99				

Operating the gas stripping column at G/L ratios between approximately 10^2 - 10^4 hydrocarbons ($K_{g/w}$ > 0.1) and even alcohols ($K_{g/w} \geqslant 10^{-3}$) can be separated from the solution during one equilibration stage.

An example of a practical application of selective enrichment with this column published some months ago is the identification of off-odour in polyethylene containing food packaging materials (20). Previous attempts with unselective separation techniques, that means extractions of all volatiles from the sample failed because harmless hydrocarbons overlapped traces of sensoric active unsaturated carbonyls, several orders of magnitude are lower in concentration than the hydrocarbons.

References

1. Swinnerton, J.W., Linnenbom, V.J., Cheek, C.H.: Anal. Chem. 34, 483 (1962).

2. Grob, K.: J. Chromatogr. 84, 255 (1973).

3. Colenutt, B.A., Thorburn, S.: Int. J. Environm. Anal. Chem. 7, 231 (1980).

4. Boren, H., Grimvall, A., Sävenhed, R.: J. Chromatogr. 252, 139 (1982).

5. Burnett, M.G.: Anal. Chem. 35, 1567 (1963).

6. Buttery, R.G., Ling, L.C., Guadagni, D.G.: J. Agr. Food Chem. 17 385 (1969).

7. Buttery, R.G., Bomben, J.L., Guadagni, D.G., Ling, L.C.: J. Agr. Food Chem. 19, 1045 (1971).

8. Vitenberg, A.G., Joffe, B.V., Dimitrova, Z.S., Butaera, J.L.: J. Chromatogr. 112, 319 (1975).

9. Kieckbusch, T.G., King, C.J.: J. Chromatogr. Sci. <u>17</u>, 273 (1979).

10. Mazza, G.: J. Food Technol <u>15</u>, 35 (1980).

11. Leighton, D.T., Jr., Calo, J.M.: J. Chem. Eng. Data <u>26</u>, 382 (1981).

12. Park, T., Rettich, T.R., Battino, R., Peterson, D., Wilhelm, E.: J. Chem. Eng. Data <u>27</u>, 324 (1982).

13. Drozd, J., Vejrosta, J., Novak, J., Jönsson, J.A.: J. Chromatogr. <u>245</u>, 185 (1982).

14. Mackay, D., Bobra, A., Shiu, W.Y.: Chemosphere <u>9</u>, 701 (1980).

15. Tewarl. Y.B., Mill r, M.M., Wasik, S.P., Martire, D.E.: J. Chem. Eng. Data <u>27</u>, 451 (1982) and May, W.E., Wasik, S.P., Freeman, D.H.: Anal. Chem. <u>50</u>, 175 (1978).

16. Pierotti, G.J., Deal, C.H., Derr, E.L.: Ind. and Eng. Chem. 51, 95 (1959).

17. Evans, M.B.: Chromatographia <u>2</u>, 397 (1969).

18. Piringer, O., Jalobeanu, M., Stanescu, U.: J. Chromatogr. <u>119</u>, 423 (1976).

19. Jennings, W., Shibamoto, T.: Qualitative Analysis of Flavor and Fragrance Volatiles by Glass Capillary Gas Chromatography, Academic Press, New York, London, 1980.

20. Koszinowski, J., Piringer, O.: Dtsch. Lebensm. Rdsch. <u>79</u>, 179 (1983).

ANALYTICAL TECHNIQUES

STATE-OF-THE-ART FUSED SILICA CAPILLARY GAS CHROMATOGRAPHY:
FLAVOR PROBLEM APPLICATIONS

Walter Jennings, Gary Takeoka

Department of Food Science and Technology
University of California, Davis CA 95616

Introduction

It is widely accepted that aroma volatiles exert major effects on flavor,
and certainly gas chromatography remains our most powerful technique for
the separation of volatile compounds. Power, however, is not an unmixed
blessing; power corrupts. It has been pointed out (1) that because of
this tremendous power, a vast multitude of inadequately trained investigators, misusing poorly designed equipment, generate a large amount of
inferior but still useful data. Our purpose today is the exploration of
how much better we could do by rectifying a few sins of commission and
sins of omission.

To a large degree, the flavor chemist has had a major responsiblity for
many of the developments in gas chromatography. This occurred because
the existing methodology was often inadequate to flavor applications,
which on many occasions entails a more demanding set of challenges than
those posed by many other systems. Many flavor mixtures are exceedingly
complex, and it is quite probable that some naturally occurring essential
oils have never yet been completely resolved. It is well recognized that
minor sample constituents can exercise effects on the sensory attributes
that may be subtle or major. While massive sample injections help ensure
that a detectable quantity of a given component is chromatographed, the
massivity of the injection interferes with component resolution. These
complications may render difficult or impossible the correlation of
important but minor components with the organoleptic properties of the
product.

An additional problem arises with the fact that some sample components are reactive, and others are relatively labile. As a consequence, some injection modes may engender compositional changes in the sample even before the chromatographic process begins.

The "sniff port" has been a favorite tool of some flavor chemists, because it permits the assessment of the level of sensory import attributable to given regions of the chromatogram. Less important regions can then be neglected, and attention focused on those regions of greater organoleptic significance. This approach requires either the use of a non-destructive detector, effluent stream splitting, or sequential separations. None of these are ideal solutions: the non-destructive detectors currently available are not truly compatible with high resolution columns; the split ratio of a stream splitter often varies with temperature; sequential separations, in which the nose is substitued in one run for the chromatographic detector in the other presupposes a constancy of solute behavior that may be altered by differences in outlet resistance and other variables whose control is problematic at best. In addition, high capacity columns (desirable for sniff testing) usually exhibit lower resolution.

In spite of these difficulties, we are able to make some very real progress, and there are developments that hold promise of even better results to come.

Injection considerations can be especially important. Good chromatographic results require not only the use of high resolution columns, but also demand a narrow sample band as it begins the chromatographic process (see, for example, reference 2). A narrow starting band can be achieved either by the injection of a narrow band per se (split injection), or by the injection of a broader band which is then subjected to narrowing or focusing before the chromatographic process commences (splitless and on-column injection). Split injections offer certain advantages, but do subject the sample to the most severe thermal shock (see, for example, reference 3), and discrimination against higher boiling solutes can be demonstrated. State-of-the-art on-column injection devices offer several advantages. The more obvious are:

1) the severe thermal shock accompanying rapid sample vaporization is eliminated; samples are vaporized only under conditions of the chromatographic process ... as soon as their vapor pressures are sufficiently high, the separation process commences;

2) higher boiling solutes that resist analysis by split or splitless injection because they cannot be volatilized with the requisite speed can be deposited directly on-column.

It soon becomes apparent, however, that our understanding of just what happens during the injection process is still imperfect; recent efforts have led to the introduction of new terms such as "band broadening in time", and "band broadening in space" (4), the "retention gap" (5), and "phase soaking" (6). In some cases, new injection procedures have been suggested that would appear to contradict previous recommendations; in only a few cases, however, are there attempts to correct conflicts in the existing recommendations (e.g. 2).

Like most practitioners, the flavor chemist is more usually interested in practical applications of gas chromatography than in the process itself. The conflicting recommendations issued by experts in the field, combined with the special hardware requirements of on-column injection, leave him frustrated and uncertain. But because of the advantages cited above, on-column injections should hold considerable appeal for the flavor chemist. Figure 1 shows a retro-fit on-column injector that utilizes two proper-pies of the fused silica column - flexibility, and low thermal mass - to permit on-column injections of a sample contained in the much more inert fused silica needle (vide infra). This injector design can be employed for the introduction of small (one to two uL), larger (five to ten uL), and even massive (e.g. 100 uL; see reference 2) samples, usually diluted in solvent, under conditions that obviate the utility of a "retention gap" (5), and avoid the "reverse thermal focus" inherent in the slow

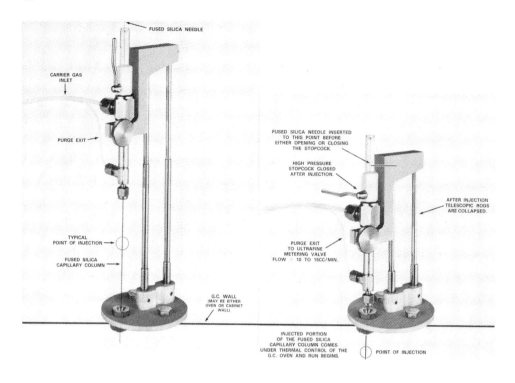

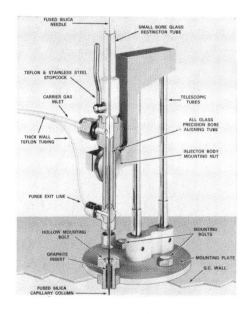

Figure 1. A retro-fit on-column injector, capable of introducing small, medium, or large amounts of liquid, solid (dissolved in solvent), or vapor samples directly in columns as small as 0.25 mm I.D., via a highly inert fused silica needle. The design, by which a predetermined length of the low-thermal mass fused silica column is withdrawn from the oven prior to injection (top left), and lowered back into the oven following injection (top right), offers many advantages. The bottom figure illustrates a cut-away view, exposing the glass alignment device guiding the fused silica needle into the fused silica column (see text for details).

ascent of a positive temperature ramp (7). By adjusting the length of the needle (i.e., the point of injection) relative to the length of column withdrawn from the oven prior to injection (the ideal values of these variables are influenced by a number of factors, including the temperature of the oven), the injection is placed on a portion of the column whose temperature is below the boiling points of all constituents in the solvent-solute plug. Within a very short distance (one to two cm), the plug (which is moving under the impetus of the carrier gas) should encounter a section of column slightly above the boiling point of the solvent, but well below the boiling point of the solutes. The solvent "distills" from the plug, leaving the solutes cold-trapped on a short length of column; this is then abruptly "stepped" to the higher oven temperature by lowering the column into the oven.

Another advantage to be gained from the above procedure lies in the inertness of the fused silica needle. Several years were required for the flavor chemist to move from metal to glass columns, and still longer before an appreciable fraction of work employed glass capillary columns. Many such columns still exhibited high levels of activity, and the degradation of sensitive compounds was not uncommon. With an increased understanding of the causes of that activity, great pains were taken to remove even traces of metal ion from the glass surface, resulting in much more inert columns. Finally, columns were constructed from the most inert of the glasses, resulting in fused silica capillary columns. But an abundant source of metal contamination remains: the stainless steel syringe needle. Syringes fitted with steel needles often comprise a miniature fractionation system per se: what comes out of that syringe is not always what went into it. The fused silica needle permits the discriminating analyst to remove the last metal from the chromatographic system. This should permit us to detect much lower levels of sulfur- and nitrogen-containing compounds, which would be expected to be quite important to certain flavors, and whose reported occurrence sometimes seems less frequent than seems reasonable.

The 1.0 um column is eminently suitable for headspace injections; methods for injecting up to one mL of headspace directly inside a fused silica column via a fused silica needle were recently described (8). The volatiles in that headspace were trapped by immersing a portion of the column in liquid nitrogen. Breakthrough of low boiling volatiles, which then suffer a reverse thermal focus from ascending a positive temperature ramp, can occur; variables such as carrier gas flow rate, sample size and the time required for injection, and the coolant temperature and the length of column subjected to cooling, all play inter-related roles.

Gas chromatography has not been an unmixed blessing to those interested in creating aromas and flavors. The gas chromatograph is a differentiator; at best, it reveals how many components are in a mixture, and how much of each component is present. (Indeed, it is doubtful whether even this restricted goal is often attained). Aromas and flavors, on the other hand, are usually due to the integrated responses to an abundance of stimuli, and the result can be complicated by synergistic and antagonistic interactions between those stimuli. Flavorists and perfumists are often interested in establishing formulations, i.e. the mixing of ingredients to produce a specific odor response. Sniff evaluation of the headspace overlying a blend of those ingredients is an accepted route, but minor changes in formulations can exercise a profound effect on the aroma, and each such change usually requires the mixing of a new batch. One method that has been used is dipping a separate "smelling blotter" into each separate ingredient, and placing the blotters in close proximity to each other. The vapors from the various ingredients mingle, and are sniffed as a mixture. Obviously, only a restricted number of ingredients can be employed, and quantitative variations and control are not possible.

Figure 2 shows a schematic of a specialized form of olfactometer (patent pending, see reference 9): sealed containers, each partially filled with a separate ingredient, are connected in parallel to a low pressure gas source. Metering valves permit the analyst to control the volume of gas conducted through the headspace vapors present in each container. The

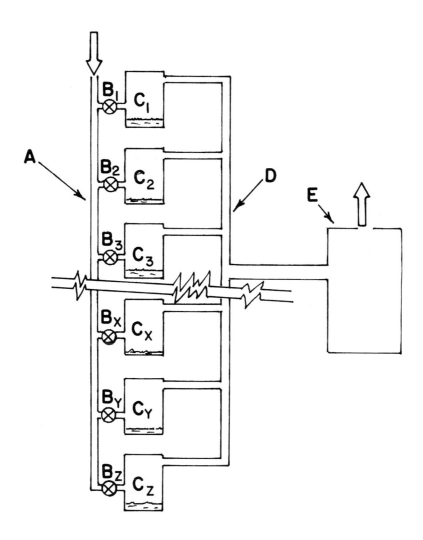

Figure 2. Schematic of a specialized olfactometer permitting the controlled blending of a number of confined headspace atmospheres, and sampling
of the overall blended headspace for organoleptic evaluation and/or instrumental analysis. The volume of low pressure gas (or air) admitted
from the low pressure manifold (A) to each of the sample chambers (C) is
controlled by the needle valve (B) connected to that chamber. The chambers can be individually thermostatted to control the vapor pressure of
the contents. The individual headspaces are conducted via the manifold
(D) to the blending and sampling chamber (E). In some cases, it may be
desirable to employ separate chambers for blending and sampling. Patent
pending. See reference (9).

headspace vapors from the various chambers are then blended by recombining the several gas streams in a central mixing chamber, which is subjected to sniff evaluation. Adjustment of the metering valves permits the headspace contribution of each ingredient to be independently varied over an infinite concentration range. The effects of such variations on the aroma impact can be quickly ascertained, and the gaseous mixture yielding a given aroma can be analyzed by headspace injection (8).

To circumvent the danger of liquid phase solubilization and phase stripping, on-column injections require the use of bonded-phase columns, which in their present form were first suggested by Jenkins and Wohleb (10), which then permitted the development of stable thick-film columns. Commercially available thick-film columns have been limited to d_f = 1.0 um, but recently Grob and Grob (11) explored the utility of "very thick films" (up to 8.0 um).

One of the major points of difference between packed and capillary (or open tubular) columns lies in the fact that the magnitude of the C term of the van Deemter equation is dictated largely by diffusivity in the stationary phase (D_s) in packed columns, and by diffusivity in the mobile phase (D_m) in capillary; in capillary columns of conventional film thickness, D_s has little or no impact on column efficiency. Figure 3 shows computer-plotted van Deemter curves (12) for d_f values of 0.1, 0.5, 1.0 and 2.0 um, D_s values differing (top vs bottom) by an order of magnitude. Two factors that are particularly important to our considerations emerge:

1) as long as D_s is sufficiently large, d_f is unimportant until values in excess of ca. 0.5 um are encountered; considerable efficiency is sacrificed at d_f values in excess of 1.0 um. This is to be expected, as the term d_f is squared in the van Deemter equation and in modifications thereof (e.g. (12-17);

2) the loss in efficiency with increasing d_f becomes magnified as D_s decreases.

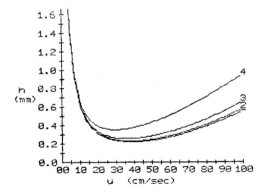

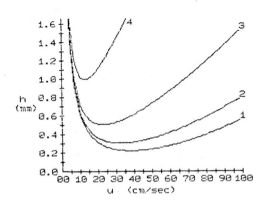

Figure 3. Computer plotted van Deemter curves (12) for a solute k = 5.0, on a .25 mm x 30 m columns with d_f values of 1) 0.1 um; 2) 0.5 um; 3) 1.0 um; 4) 2.0 um. Diffusivities in the mobile phase (D_m) assumed constant at .35 cm^2sec^{-1}. Diffusivities in stationary phase (Ds) assumed at (top) 9.3 x 10^{-6}, and (bottom) 9.3 x 10^{-7}. See text for discussion.

This latter observation might hold some ominous premonitions for bonded phases, as these would normally be expected to possess greater viscosities (and lower diffusivities) than their unbonded counterparts. However, data of Hawkes et al. (18-20) established that diffusivities were higher in SE 30, a high viscosity dimethyl polymethyl siloxane gum, than in OV-101, a low-viscosity dimethyl polymethyl siloxane fluid (mol. wt. 4 x 10^5 vs 1.1 x 10^4). They suggested that the gum film possessed a more porous, open structure than did the fluid film. Increased cross-linking leads to more highly viscous forms of the liquid phase, whose structures are perhaps even more porous. This would help to explain why the lower temperature limit (short term) for cross-linked, bonded forms of dimethyl polymethyl siloxane (e.g. DB-1) are around -60°C, as compared to lower temperature limits of +50°C for the non-cross-linked non-bonded counterpart (e.g. SE-30, Supelco listing). It is obvious, however, that D_s will become critically important as d_f increases.

Grob and Grob (11) evaluated their separation efficiencies on thick film columns in terms of separation numbers, TZ, generated under programmed conditions. These results are informative as employed, but they can be misleading, as several inter-related factors may tend to become confused. The obvious benefit that increases in d_f exercise on component resolution results from the fact that the isothermal partition ratios, k, vary directly with d_f:

$$K_D = \beta k = \frac{r}{2d_f}\, k;$$

$$k = \frac{d_f 2 K_D}{r}$$

Both K_D and k are continuously changing during temperature programming; separation numbers are very much a function of the average k of the test solute, and are also affected by temperature, even at constant k (21). One effect of the very thick film is to cause a massive shift in k (isothermal), which results in a significant increase in TZ.

The number of theoretical plates required to separate any two solutes can be estimated from the relationship (see, for example, reference 3):

$$n_{req} = 16Rs^2 \left(\frac{k+1}{k}\right)^2 \left(\frac{\alpha}{\alpha-1}\right)^2$$

Under isothermal conditions, thicker films will enhance solute resolution until the decreases in the values of the multipliers $[(k+1)/k]^2$ resulting from the thicker film no longer compensate for the increased values of h which also result from the thicker film.

For a solute k = 0.1, the value of this multiplier is 121; if d_f is increased by a factor of 20, k becomes 2.0 and the value of the multiplier drops to 2.25. All other things being constant, a separation that required (e.g.) 2,688,000 theoretical plates in the first case (k = 0.1) could be achieved with 50,000 theoretical plates in the second case (k = 2); i.e. we would need approximately 1/50 as many theoretical plates. As long as the increase in h occasioned by the increase in d_f does not exceed a factor of 50, improved separation will be achieved on this low-k solute.

For a solute k = 1, the value of the multiplier is 4; increasing d_f by that same factor of 20 will increase k (isothermal) to 20, and the multiplier becomes 1.1. In this case, increasing h even by a factor of 4 (all other things being equal) will have an adverse affect on separation. These inter-relationships are further complicated by the fact that because a given solute is retained longer by the thicker film (k is larger), it is, in programmed operation, subjected to higher temperatures in the course of that run; higher temperatures result in lower values for K_D and k, and higher values for D_s and D_m. Film thicknesses of up to 1.0 um are generally satisfactory for all but the highest boiling or most thermally labile solutes, unless that phase abnormally restricts solute diffusivities; thicker film columns are usually useful only in special situations. The 1.0 um column is eminently suitable for headspace injections.

I would like to conclude on a somewhat different note. Most flavor chemists are what might be termed "practical chromatographers", more interested in analytical results than in esoteric theoretical considerations. Today's world, however, has become considerably more complicated for the pragmatic investigator. He (or she) is faced with

1) a bewildering choice of columns that vary in length, diameter, liquid phase film thickness, type of liquid phase, suitability for different samples, tendency to "bleed" or degrade, the quality of the glass or fused silica, and the efficiency of the coating process;

2) decisions relative to operating parameters, including the choice of the carrier gas, and selection from arrays of selectable temperatures, program conditions, and carrier gas flows that are essentially infinite and closely inter-related;

74

3) sample- and goal-dependent decisions as to the most suitable means of sample injection and solute detection.

The qualitative and quantitative validity of chromatographic results, and the conclusions based thereon, can be drastically influenced by any or all of these various parameters. Average practitioners of gas chromatography, numbered in the hundreds-of-thousands, have been so busy in their own fields of endeavor that they have failed to stay current with developments in chromatographic science. Knowledge gaps of this type, where two diverse fields of knowledge are both growing rapidly, have an unfortunate tendency to become wider rather than narrower. Particularly in our field of flavor chemistry, we have lost a degree of communication between the "practical chromatographer" and the "theoretical chromatographer"; this is regrettable, and that lapse must be rectified if we are to reap maximum benefits from this rapidly advancing field. The Foreward of an early classic in gas chromatography (22) contains observations which, far from fading in significance, seem even more pertinent today: "Finally, I cannot resist an attempt to destroy the blind, artificial barrier inserted by some between theory and practice. Theory (when correct) and experiment (if carefully executed) describe the same truths. The science of chromatography requires both approaches if it is to grow in proportion to the demands made on it".

References

1. Jennings, W. G.: Comparisons of Fused Silica and Other Glass Columns in Gas Chromatography, Huethig Verlag, Heidelberg-Basel-New York 1981

2. Jenkins, R., Jennings, W.: HRC & CC 6, 228-231 (1983).

3. Jennings, W.: Gas Chromatography with Glass Capillary Columns, second edition, Academic Press, New York-San Francisco-London 1980

4. Grob, K., Jr.: J. Chromatogr. 213, 3-14 (1981).

5. Grob, K., Jr., Mueller, R.: J. Chromatogr. 244, 185-196 (1982).

6. Grob, K., Jr., Schilling, B.: J. Chromatogr. 259, 37-48 (1983).

7. Jennings, W., Takeoka, G.: Chromatographia 15, 575-576 (1982).

8. Jennings, W. Proceedings of the IFT symposium on Modern Methods of Food Analysis, 10-11 June 1983, New Orleans LA. Avi Publishing (in

press).

9. Litman, I. 41 Holiday Park Dr., Hauppauge NY 11788. Personal communication, July, 1983.

10. Jenkins, R. G., Wohleb, R. H.: paper presented at the 15th International Symposium "Advances in Chromatography", October 06-09, Houston TX 1980.

11. Grob, K., Grob, G.: HRC & CC 6, 133-139 (1983).

12. Ingraham, D. F., Shoemaker, C. F., Jennings, W.: HRC & CC 5, 227-235 (1982).

13. Giddings, J. C., Seager, S. L., Stucki, L. R., Stewart, G. H.: Anal. Chem. 32, 867-870 (1960).

14. Giddings, J. C.: Anal. Chem. 34, 314-319 (1962).

15. Giddings, J. C.: Anal. Chem. 36, 741-744 (1964).

16. Sternberg, J. C., Poulson, R. E.: Anal. Chem. 36, 58-63 (1964).

17. Cramers, C. A., Wijnheymer, F. A., Rijks, J. A.: HRC & CC 2, 329-334 (1979).

18. Butler, L., Hawkes, S. J.: J. Chromtogr. Sci. 10, 518-523 (1972).

19. Kong, J. M., Hawkes, S. J.: J. Chromatogr. Sci. 14, 279-285 (1976).

20. Millen, W., Hawkes, S. J.: J. Chromatogr. Sci. 15, 148-150 (1977).

21. Jennings, W., Yabumoto, K.: HRC & CC 3, 177-179 (1980).

22. Giddings, J. C.: Dynamics of Chromatography, Marcel Dekker, New York 1965, page viii.

NEW DEVELOPMENTS IN ENANTIOMER SEPARATION BY CAPILLARY
GAS CHROMATOGRAPHY

Wilfried A. König
Institut für Organische Chemie und Biochemie der Universität
Hamburg, D-2000 Hamburg 13, F.R.G.

Introduction

The majority of molecular interactions in biochemistry is regulated by enzymes and proceeds strictly stereospecifically.
Also the receptors for flavor and taste clearly distinguish
between substrates of different stereochemistry. The investigation of the configuration of chiral molecules has always been
an essential part of structure elucidation of natural compounds.

Capillary gas chromatography with its superior separation efficiency and sensitivity together with a wide range of selectivity of a stationary phase towards a substrate can be used to
differentiate even between stereoisomers, resulting in resolution of these isomers.

Chiral compounds may either be separated as diastereoisomers on
conventional stationary phases or as enantiomers on chiral stationary phases. In the first case chiral reagents are used to
form diastereomeric derivatives with a functional group of the
chiral substrate. In the second case the two enantiomers to be
resolved form diastereomeric association complexes with the
chiral stationary phase. These complexes differ in free energy
and consequently a separation of the enantiomers is observed.
Complex formation is brought about by different types of molecular interactions with hydrogen bonding and dipole-dipole-inter-
action being most important.

Analysis of Volatiles
© 1984 Walter de Gruyter & Co., Berlin · New York – Printed in Germany

In recent years the development of new temperature stable
chiral stationary phases and the use of new types of derivatives have extended direct enantiomer separation to an un-
expectedly large variety of chiral compound classes.

Results

1) *Low Molecular Weight Stationary Phases*. Enantiomer separation on chiral stationary phases was first observed by E.
   Gil-Av (1) and his associates. They separated volatile
   amino acid derivatives by enantioselective interaction
   with stationary phases derived of amino acids or dipep-
   tides. On stationary phases carrying an L-amino acid residue the L-enantiomer of a mixture showed a longer retention
   time than the D-enantiomer and vice versa.

   After these first reports on enantioselective gas chromatography many research groups have worked on new procedures
   to improve thermal stability and enantioselectivity and to
   extend enantiomer separation to other compound classes (2,
   3). These efforts were greatly stimulated by the development of modern glass capillary gas chromatography.

2) *Polymer Stationary Phases*. A great enhancement in thermal
   stability of chiral stationary phases was achieved by the
   application of chiral polysiloxanes by Frank, Nicholson and
   Bayer (4-7). By copolymerisation of dimethylsiloxane and
   carboxyalkyl-methylsiloxane a functionalized polymer was
   obtained with carboxylic groups in the side chains. To
   these groups L-valine-tert.butyl amide was connected in an
   amide bond. N-acylated derivatives of L-valine-tert.butyl
   amide had earlier been proposed by Feibush (8) and Gil-Av
   et al. (9, 10) as highly enantioselective stationary phases
   for amino acids and amino alcohols.

   The chiral polysiloxane chirasil-val could be used upto

above 220°C and showed excellent results in separating the
enantiomers of amino acids, some drugs of the ephedrine
type (11) and α-hydroxyacid amides (5).

3) *Chiral Polymers Derived of Cyanoalkyl Polysiloxanes.* Verzele and his coworkers (12, 13) suggested a different way
to obtain chiral polysiloxanes. Starting from commercially
available polysiloxanes with cyanoalkyl side chains carboxylic groups could be formed by acid hydrolysis of the
cyano groups and by covalently binding amino acid derivatives. The results obtained with modified polysiloxanes
OV-225 and Silar 10C and L-valine-tert-butylamide connected
to the side chains were encouraging but not quite equivalent to chirasil-val.
The polysiloxane XE-60 was modified by our group and L-
valine-(S)- or (R)-α-phenylethylamide was attached to the
carboxylic groups after alkaline hydrolysis of the cyano
groups (14, 15). The use of two asymmetric centers proved
to be advantageous. As shown in numerous examples enantioselectivity depends on the chirality of both residues.
Both diastereoisomeric polymers are stable at least upto
220°C and have complementary separation properties. In a
routine stereochemical investigation a gas chromatograph
equipped with *one* injection port and *two* detectors would
serve as an instrument for tandem operation of two equivalent capillaries with the two diastereoisomeric stationary
phases and with simultaneous injection of a sample on both
columns (16).
In a different approach XE-60 and similar polysiloxanes
were modified by reduction of the cyano groups to amino-
methyl groups with $LiAlH_4$ (17). Amino acid derivatives
with free carboxylic groups could be connected to the polymer. These stationary phases also show interesting separation properties. However, the scope of this type of chiral
polymers is not sufficiently investigated yet.

80

4) *Enantiomer Separations on XE-60-L-valine-(S)-α-phenylethyl-
amide (S-phase) and XE-60-L-valine-(R)-α-phenylethylamide
(R-phase).* As already mentioned these stationary phases
as diastereoisomers exhibit different enantioselectivities
towards many chiral compounds. The S-phase is particularly
suited to separate amino acids (14) as demonstrated in
fig. 1.

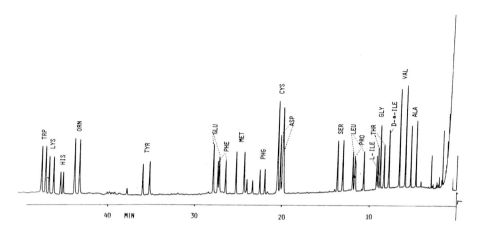

Figure 1   Trifluoroacetyl-DL-amino acid isopropyl esters.
15 m fused silica capillary with XE-60-L-val-(S)-α-phenyl-
ethylamide (Chrompack); 70°C - 180°C, 2.5°C/min.

The identification of the configuration of amino acids in
peptide antibiotics is an important part of structure elucidation of this group of natural products. Also very precise determinations of the ratio of racemization during
peptide synthesis as well as the control of optical purity
of amino acids used for synthesis of biologically active
peptides are preferentially performed with this gas chromatographic method. The R-phase displays a unique enantioselectivity for chiral amines (19) (fig. 2) and amino alcohols (15). Valinol, leucinol and phenylalaninol have been
found as constituents of some membrane modifying peptide
antibiotics (19). We also found that it is possible to

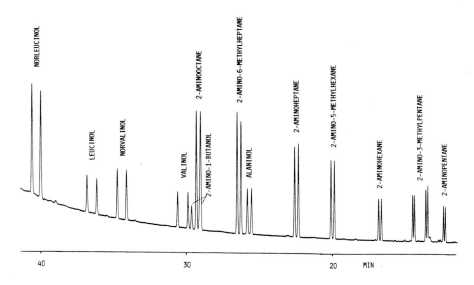

Figure 2   Trifluoroacetyl derivatives of chiral amines and
amino alcohols. 50 m fused silica capillary with XE-60-S-val-
(R)-α-phenylethylamide (Chrompack); 100°C - 180°C, 1.5°C/min.

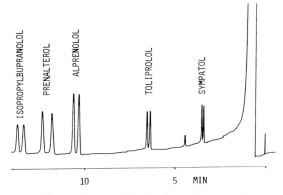

Figure 3   Heptafluorobutyryl derivatives of some racemic
drugs. 18 m glass capillary with XE-60-L-val-(R)-α-phenyl-
ethylamide; 150°C.

separate the enantiomers of amino alcohols of the ephedrine
type and ß-adrenoceptor drugs (ß-blockers) as heptafluorobutyryl derivatives (20). An example is given in fig. 3.

Although hydrogen bonding association was considered for
a long time to be essential for enantiomer separation,
several cases of "chiral recognition" have been observed
where hydrogen bonding is of minor importance or not possible at all (21-23). So it was quite unexpected that the
enantiomers of carbohydrates, methyl glycosides (15, 24,
25) and polyols (18) can be separated on either the S- or
the R-phase, without a systematic preference in enantioselectivity for one or the other configuration of the
phenylethylamide residue. The separation of D- and L-
galactose derivatives was used to determine the ratio of
these enantiomers in snail galactans (24).

5) *New Derivatization procedures for Enantioselective Gas
    Chromatography.* Enantioselectivity is the result of ste-
    reospecific molecular interactions of the enantiomers of
    a sample to be resolved and a chiral stationary phase. It
    therefore depends on structural features of both interac-
    ting partners. With the availability of thermostable poly-
    mers as chiral stationary phases it has become possible to
    alter the type of derivatives since volatility became less
    important. The higher operating temperatures of the capil-
    lary columns allow the use of less volatile and more polar
    derivatives. The attempts to separate chiral alcohols in
    free form or as acylated derivatives were unsuccessful ,
    except for only a few examples (26). A general way for the
    separation of alcohol enantiomers was found after formation
    of urethane derivatives (27) (scheme 1). Isopropyl isocya-
    nate proved to be especially suited as reagent. Usually
    good yields of urethane derivatives are obtained by heating
    samples of 100 μg of alcohols in a 1:1 mixture of dichloro-
    methane and isopropyl isocyanate at 100°C for 30 - 60 min.
    A large number of chiral aliphatic and terpene alcohols
    could be separated with this method. Chiral alcohols have
    been found as active constituents of insect communication
    systems (pheromones) or as flavor compounds. Usually the

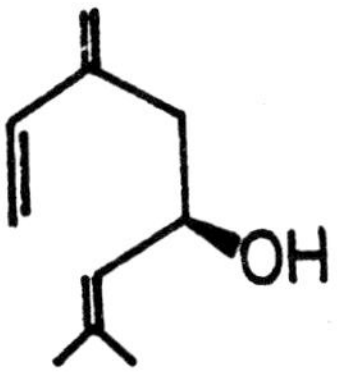

1

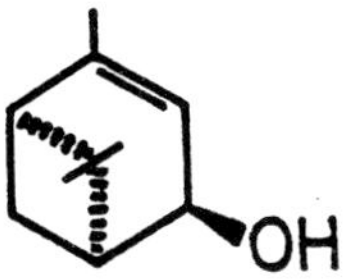

2

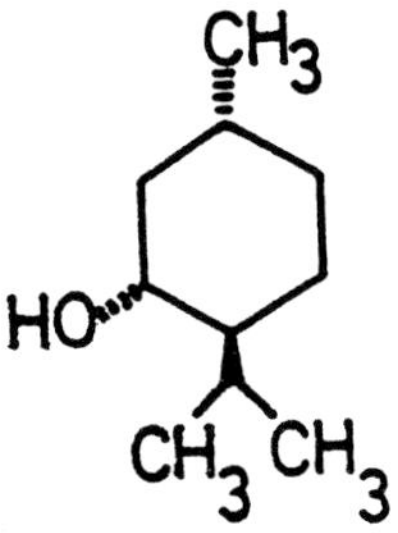

3

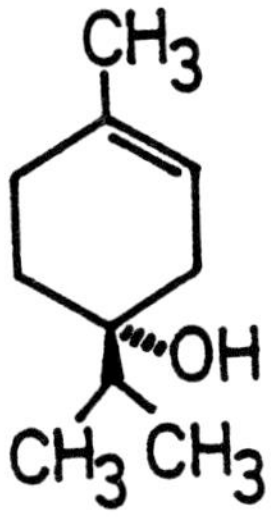

4

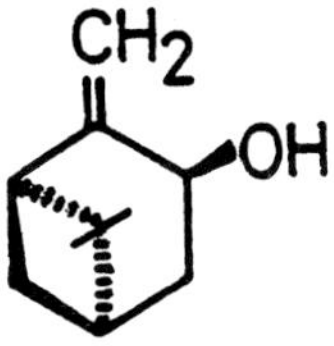

5

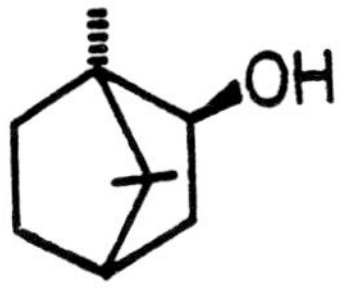

6

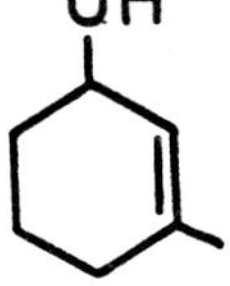

7

8

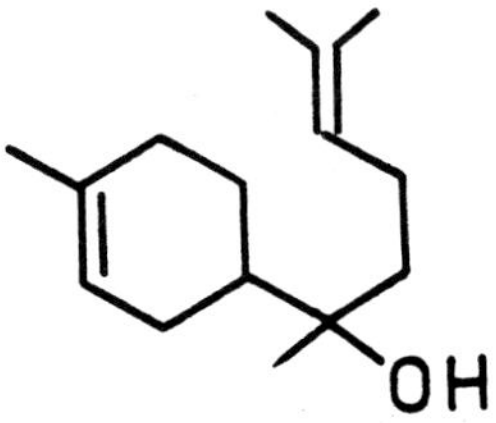

9

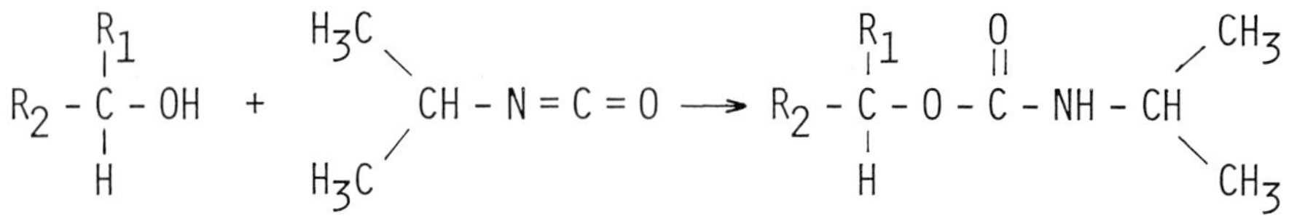

$$R_2 - \overset{\overset{\displaystyle R_1}{|}}{\underset{\underset{\displaystyle H}{|}}{C}} - OH \quad + \quad \overset{\displaystyle H_3C}{\underset{\displaystyle H_3C}{\diagdown}} CH - N = C = 0 \quad \longrightarrow \quad R_2 - \overset{\overset{\displaystyle R_1}{|}}{\underset{\underset{\displaystyle H}{|}}{C}} - 0 - \overset{\overset{\displaystyle 0}{||}}{C} - NH - CH \overset{\displaystyle \diagup CH_3}{\underset{\displaystyle \diagdown CH_3}{}}$$

Scheme 1

biological activity is connected to a certain configuration. Some of the alcohols separated are 2-heptanol and 3-octanol (alarm pheromones of ants), 4-methyl-3-heptanol (aggregation pheromone of the elm bark beetle), ipsdienol (1) (aggregation pheromone of certain *Ips* and *Pityokteines* bark beetles). Terpinenol-4 (4), *trans*-pinocarveol (5), *trans*-verbenol (2), 6-methyl-5-hepten-2-ol (sulcatol) (8), and 1-methyl-cyclohexene-3-ol (seudenol) (7) have also been found to be active as pheromones in several bark beetles.

Another group of chiral alcohols are known to be important flavor constituents. We have shown that all eight stereo-isomers of menthol (3) can be separated as isopropyl urethanes (28). It was also possible to separate isoborneol (6), the four stereoisomers of α-bisabolol (9) and α-ionol (fig. 4). An improvement of these separations should be possible by directly coupling capillary columns with a *non-chiral* phase for pre-separation of diastereoisomers and a *chiral* phase for subsequent enantiomer separation.

In some cases it was not possible to separate the enantio-mers of a chiral alcohol, for instance in the case of 1-octen-3-ol, the major flavor constituent of mushrooms. Nevertheless a determination of the configuration is possible after catalytical micro-hydrogenation of 1-octen-3-ol to yield octan-3-ol, which can readily be separated (fig. 5). In extending the reaction of isocyanates to other

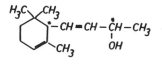

α-Ionol

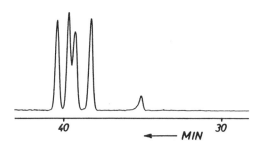

Figure 4   Separation of the 4 stereoisomers of α-ionol as
isopropyl urethane derivatives. 40 m glass capillary with
XE-60-L-val-(S)-α-phenylethylamide; 150°C.

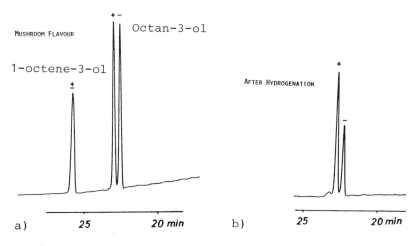

Figure 5   Gas chromatogram of isopropyl urethane derivatives
of a) racemic 1-octene-3-ol and octane-3-ol and b) of (+)-1-
octene-3-ol and racemic octane-3-ol after micro-hydrogenation.
Same column as in fig. 4; 130°C.

compound classes we showed that α-hydroxy acid esters (29) can also be separated very nicely after formation of urethane derivatives. As in the case of amino acids on a stationary phase containing L-valine the L-enantiomers of α-hydroxy acids have longer retention times than their antipodes.

Isocyanates also readily react with chiral amines to form ureido derivatives. This reaction already proceeds at room temperature. The products can be separated with high separation factors (29). Similarly α-amino acids can be separated after esterification and reaction with isocyanates at room temperature in the presence of a trace of pyridine. tert.Butyl isocyanate proved to be superior to isopropyl isocyanate (30). This type of derivative may be advantageous over trifluoroacetyl derivatives for amino acids like proline and pipecolic acid (fig. 6). These amino acids usually can not be completely separated as TFA-derivatives, even on very long capillaries.

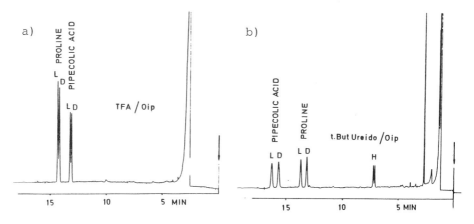

Figure 6   Separation of DL-proline and DL-pipecolic acid a) as N-trifluoroacetyl/isopropyl esters on a 50 m fused silica capillary with XE-60-L-val-(S)-α-phenylethylamide (Chrompack) at 120°C and b) as tert.butylureido/isopropyl esters on a 25 m glass capillary with XE-60-L-val-(S)-α-phenylethylamide at 150°C. (H = hydantoin of pipecolic acid)

In trying to separate the urethane derivatives of ß-hydroxy
acid esters one finds that their separation is very poor
and in no case a baseline separation could be achieved.
The same problem is observed with the ureido derivatives
of N-methylamino acid esters. However, when the free ß-
hydroxy acids are heated in a mixture of methylene chloride and isocyanate a reaction takes place at the hydroxy
group *and* at the carboxylic group (28, 30). In one reaction step both an urethane and an amide group are formed
adding a second CO-NH-function to the derivative (scheme 2).

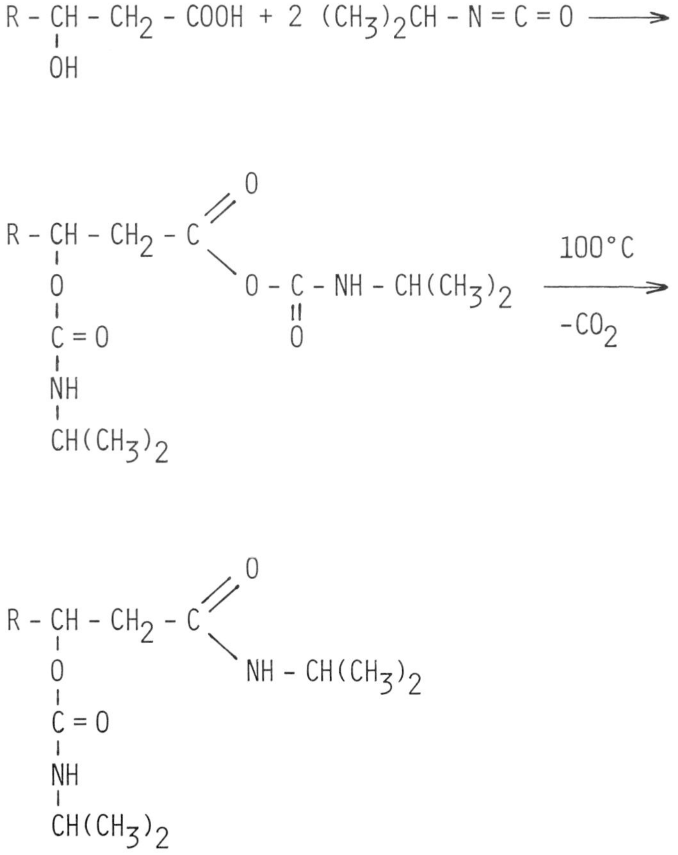

$$R - CH - CH_2 - COOH + 2\ (CH_3)_2CH - N = C = O \longrightarrow$$
$$|$$
$$OH$$

$$R - CH - CH_2 - C \overset{O}{\diagup} \quad \quad 100°C$$
$$| \quad \quad \quad \diagdown O - C - NH - CH(CH_3)_2 \longrightarrow$$
$$O \quad \quad \quad \quad \| \quad \quad \quad \quad -CO_2$$
$$C = 0 \quad \quad \quad \quad O$$
$$|$$
$$NH$$
$$CH(CH_3)_2$$

$$R - CH - CH_2 - C \overset{O}{\diagup}$$
$$| \quad \quad \quad \diagdown NH - CH(CH_3)_2$$
$$O$$
$$C = 0$$
$$|$$
$$NH$$
$$CH(CH_3)_2$$

Scheme 2

Only in this way sufficient sites for enantioselective interaction with the chiral stationary phase are supplied
and resolution of enantiomers is possible. In an analogous
way N-methylamino acids are converted to N-ureido- and
amide derivatives. Again this is the only way to separate
the enantiomers of N-methylamino acids (30), which are
common constituents of peptide antibiotics and play an
important role in modern peptide synthesis.

Chiral α-branched carboxylic acids can also be converted
into isopropyl amides by reaction with isopropyl isocyanate and be separated into enantiomers (28).

The compounds dealt with so far have all either amino,
hydroxy or carboxylic groups. All these groups are con-
vertible into derivatives with CO-NH-groups, which are
suited to form hydrogen bonded association complexes.

For chiral ketones formation of derivatives is also necessary since free ketones are not separated into enantiomers. Out of the various types of derivatives only the
oximes could be separated (31). Apparently the N-OH-group
is sufficiently polar for complex formation with the stationary phase. As soon as the OH-group is substituted by
a methyl or trimethylsilyl group separation is lost.

Except for their importance in flavor chemistry chiral
ketones were found to be alarm pheromones of ants (32-34)
and wasps (35). As an example figure 7 shows a separation
of camphor enantiomers.

Conclusions

In concluding it should be emphazised that enantioselective
gas chromatography has proved to be a sensitive analytical
tool for stereochemical investigations not only for the analysis of natural compounds but also for product control of

microbial or enzymatic biotransformations for pharmacokinetic
studies of chiral drugs and for proving the stereoselectivity
of asymmetric synthesis, which plays an increasingly important
role in modern organic chemistry.

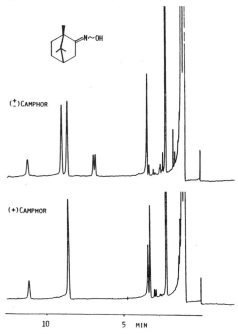

Figure 7   Separation of oxime derivatives of racemic camphor
and comparison with natural (+)-camphor. 25 m fused silica
capillary with XE-60-L-val-(S)-α-phenylethylamide (Chrompack).
140°C.

References

1. Gil-Av, E., Feibush, B., Charles-Sigler, R.: in A.B.
    Littlewood (Editor), "Gas Chromatography 1966", Institute
    of Petroleum, London (1966), p. 227.

2. König, W.A.: HRC & CC 5, 588 (1982).

3. Liu, R.H., Ku, W.W.: J. Chromatogr. 271, 309 (1983).

4. Frank, H., Nicholson, G.J., Bayer, E.: J. Chromatogr. Sci.
    15, 174 (1977).

5. Frank, H., Nicholson, G.J., Bayer, E.: Angew. Chem. 90,
    396 (1978); Angew. Chem. Int. Ed. Engl. 17, 363 (1978).

6. Frank, H., Nicholson, G.J., Bayer, E.: J. Chromatogr. 167, 187 (1978).

7. Nicholson, G.J., Frank, H., Bayer, E.: HRC & CC 2, 411 (1979).

8. Feibush, B.: J. Chem. Soc. Chem. Commun. 1971, 544.

9. Feibush, B., Balan, A., Altman, B., Gil-Av, E.: J. Chem. Soc. Perkin II 1973, 2094.

10. Charles, R., Gil-Av, E.: J. Gas Chromatogr. 5, 257 (1980).

11. Frank, H., Nicholson, G.J., Bayer, E.: J. Chromatogr. Biomed. Appl. 146, 197 (1978).

12. Saeed, T., Sandra, P., Verzele, M.: J. Chromatogr. 186, 611 (1979).

13. Saeed, T., Sandra, P., Verzele, M.: HRC & CC 3, 35 (1980).

14. König, W.A., Sievers, S., Benecke, I.: in R.E. Kaiser (Editor), "Proceedings of the IVth International Symposium on Capillary Chromatography", Institute for Chromatography, Bad Dürkheim, and Dr. A. Hüthig Publishers, Heidelberg (1981), p. 703.

15. König, W.A., Benecke, I., Sievers, S.: J. Chromatogr. 217, 71 (1981).

16. König, W.A., unpublished results.

17. König, W.A., Benecke, I.: J. Chromatogr. 209, 91 (1981).

18. König, W.A., Benecke, I.: J. Chromatogr. (1983), in press.

19. König, W.A., Aydin, M.: in "Chemistry of Peptides and Proteins", W. Voelter, E. Wünsch, J. Ovchinnikov and V. Ivanov, eds., Walter de Gruyter & Co, Berlin, New York, 1982, p. 173.

20. König, W.A., Ernst, K.: J. Chromatogr., in press.

21. Stölting, K., König, W.A.: Chromatographia 9, 331 (1976).

22. König, W.A., Sievers, S.: J. Chromatogr. 200, 189 (1980).

23. König, W.A., Sievers, S., Schulze, U.: Angew. Chem. 92, 935 (1980); Angew. Chem. Int. Ed. Engl. 19, 910 (1980).

24. König, W.A., Benecke, I., Bretting, H.: Angew. Chem. 93, 688 (1981); Angew. Chem. Int. Ed. Engl. 20, 693 (1981).

25. Benecke, I., Schmidt, E., König, W.A.: HRC & CC 4, 553 (1981).

26. Oi, N., Kitahara, H., Inda, Y., Doi, T.: J. Chromatogr. 213, 137 (1981).

27. König, W.A., Francke, W., Benecke, I.: J. Chromatogr. 239, 227 (1982).

28. Benecke, I., König, W.A.: Angew. Chem. 94, 709 (1982);
    Angew. Chem. Int. Ed. Engl. 21, 709 (1982); Angew. Chem.
    Suppl. (1982) 1605 - 1613.

29. König, W.A., Benecke, I., Sievers, S.: J. Chromatogr.
    238, 427 (1982).

30. König, W.A., Benecke, I., Lucht, N., Schmidt, E., Schul-
    ze, J., Sievers, S.: Proceed. 5th Intern. Symp. Capillary
    Chromatogr., Riva del Garda, 1983, J. Rijks, ed., Else-
    vier Scientific Publishing Company, Amsterdam, 1983,
    p. 609.

31. König, W.A., Benecke, I., Ernst, K.: J. Chromatogr. 253,
    267 (1982).

32. McGurk, D.J., Frost, J., Eisenbraun, E.J., Vick, K.,
    Drew, W.A., Young, J.: J. Insect Physiol. 12, 1435 (1966).

33. Riley, R.G., Silverstein, R.M., Moser, J.C.: Science 183,
    760 (1974).

34. Fales, H.M., Blum, M.S., Crewe, R.M., Brand, J.M.: J. In-
    sect Physiol. 18, 1077 (1972).

35. Fales, H.M., Jaouni, T.M., Schmidt, J.O., Blum, M.S.: J.
    Chem. Ecol. 6, 895 (1980).

COMPARISON OF GLC CAPILLARIES

W. Günther, K. Klöckner
WGA, Düsseldorf

F. Schlegelmilch and S. Roukeria
FH Niederrhein, Krefeld

Since some years the importance of capillary columns in gas chromatographic (GLC) studies is more and more increasing. For this reason, a comparison of GLC capillaries will be provided mainly discussing the influence of column wall pretreatments. Because of the high separation capacity of capillary columns, the selectivity of stationary phases is not of the same importance as if packed columns are used. In general, not more than two liquid stationary phases are needed for more than 90 % of normal common separations, i.e. (1) a high temperature resistant apolar dimethylpolysilane such as OV-1, and (2) a high temperature resistant polar phase such as a polyether with esterified end groups, e.g. WG-11 or OV 351.

Generally, there are two different types of capillaries, WCOT and SCOT capillaries. In the following, especially WCOT columns will be discussed. Due to their importance it will be firstly referred to the Al_2O_3- and Molsieve-13 X-coated capillaries. The Al_2O_3-coated capillaries are nearly ideal capillaries for separating low boiling hydrocarbons, especially when live-chromatography is used. They are also important for the separation of naphtenes from saturated hydrocarbons as necessary in "PIANO-Analyzing" (paraffins, isoparaffins, aromatics, naphtenes, olefins).

For SiO$_2$-capillaries, the following column materials are useful:

Fig. 1 Column materials

Soda lime glass, borosilicate glass, and fused silica can work
together. Because of the low content of kations, fused silica
should be used instead of fused quartz. Normally, fused silica
has an amount of kations less than 0.16 ppm. Pyrex is more
effective than duran in handling pretreatment. The best features
of soft glass are found in sodium calcium glass. The following
figure shows the composition of the column wall materials we
used:

	SiO$_2$	B$_2$O$_3$	Al$_2$O$_3$	Na$_2$O
Soda lime glass	70%	2%	5%	15%
Borosilicate glass	80%	12%	3%	3%
Fused silica	> 99,95%	K<0.05ppm		Na < 0,01ppm

-OH-Groups	1000 ppm for permanent bonding	of stat.
	5 ppm for normal coating	liquid
		phase

Fig. 2 Impurities in soda lime glass, borosilicate glass
        and quartz

The brittleness increases from soft glass via borosilicate glass to fused silica. Therefore, it is necessary to coat the fused silica with a lacquer to protect it against mechanical scratching. It must be pointed out that the lacquer is not ruptured in order to avoid fractures, i.e. carrier gas may leak as the fused silica could have been broken as a result of bending. Therefore, we would like to point out again that fused silica has the highest brittleness and that just the lacquer prevents a complete break-down into small pieces (as it is often demonstrated by Scanning Electron Microscopes). For varnishing polyimide is useful. Vapour metallizing with aluminium is not acceptable because of the diffusity of aluminium which may destroy the liquid stationary phase. The durability of the column is decreased, and drifts when running temperature programmes may increase. For mechanical protection, it is normally necessary to varnish up to five times with a polyimide lacquer. The next two figures show a fused silica capillary thinner coated than the thickness of fused silica wall (15):

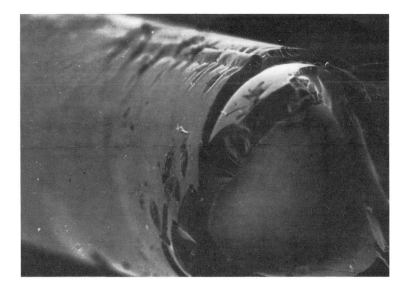

Fig. 3   Quartz with a 2 μm polyimide lacquer

Fig. 4   Quartz with a 20 µm polyimide lacquer

In the following figure 5, the polyimide rupture and the fused silica fracture are shown.

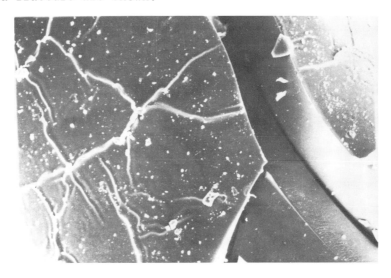

Fig. 5   Polyimide ruptures

To get similar results for fused silica column walls with borosilicate glass, it is necessary to remove the kations. This is
possible by using (1) static leaching, (2) dynamic leaching.
The leaching process is possible by using aqueous HCl or just
water. Leaching by using water is basically time-consuming. The
simplest and quickest leaching can be done by static method
according to Grob. If using this method, new surfaces result
with increased brittleness. These surfaces are similar to fused
silica in the amount of kations but only in very thin layers at
the surface, i.e. bad diffusions to the surface are possible.

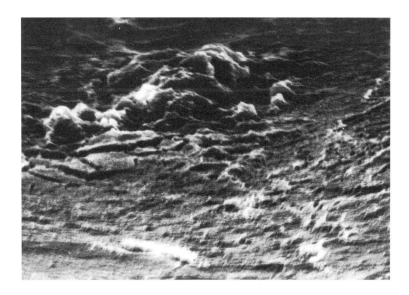

Fig. 6 Leached borosilicate glass

Soft glass should not be leached because there is no method of
removing the kations. Soft glass should be treated as described
by Alexander and Schomburg. The so-formed sodiumchloride is
left in the capillary as shown in figure 7.

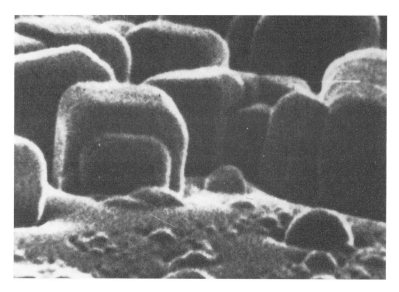

Fig. 7   NaCl crystals on soda lime glass

The bariumcarbonate method as described by Grob has no universal application.

Fig. 8 $BaCO_3$-layer

Incidentally, the waste is too large (13). The $BaCO_3$-treatment is more effective for borosilicate glass than for soft glass.

Fused silica capillaries may be used without any of the pre-described pretreatments.

All of the three column wall materials should be deactivated before coating with the liquid stationary phases, i.e. essent-tially inactivation of the silanol groups. Soft glass is a basic, borosilicate glass a slightly acidic, and fused silica an acidic material. After HCl-treatment soft glass and borosilicate glass are in acidic constitution (5). Soft glass can be deactivated by using non-ionic surfactants,i.e. Carbowax 20 M, which has slightly surfactant properties. Before deactivating these soft glass capillaries dehydration is necessary. Under the HCL con-dition by gas phase reaction in the range of about 500°C, H_2O splits under crosslinking of SiO_2-chains (6).

Fig. 9 NaCl crystals

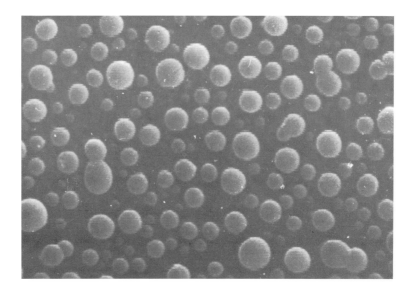

Fig. 10   NaCl crystals non-dehydrated

As seen by Scanning Electron Microscopy (SEM), the dehydration
is absolutely necessary to get illustrated angular crystals (1).

Chromatographically, the following pictures result (2,11,13,20):

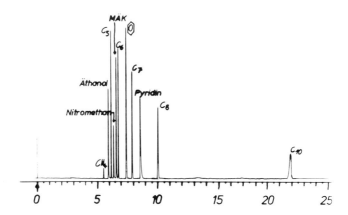

Fig. 11   Rohrschneider mixture on deactivated non-dehydrated
          NaCl-underground. Conditions: Siemens L 350; 110°C;
          WGA Splitter 0.7:0.6 bar N_2; thin-film glass capillary
          column OV 101, 40 m; desactivated; non-dehydrated.

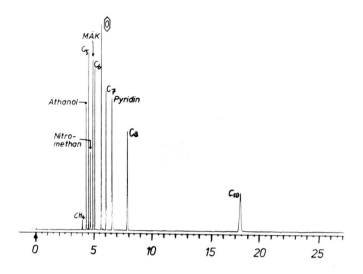

Fig. 12   Rohrschneider mixture on dehydrated and deactivated
          NaCl-underground. Conditions: cp. Fig. 11

The main quantitative differences will be seen in a remarkable
tailing decrease when using dehydrated capillaries. The disadvantage in using soft glass capillaries is the upper temperature
limit due to the lower thermal stability of the deactivation
layer. Destroying the liquid phase by kation effects is normally
not advisable as the upper limit is about +220°C and destroying
by kation effects starts at about +250°C. This is the essential
advantage of the H_2F_2/HCl-treatment.

The best way in deactivating is the Polysiloxan Degradation
method as described by Schomburg. This procedure is useful for
untreated and HCl-treated soft glass. Our studies showed that
the deactivation by the PSD-method acts through cyclic siloxanes.
The direct deactivation of soft glass by cyclic siloxanes is
impossible.

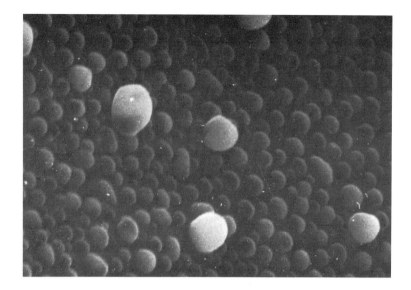

Fig. 13 NaCl-NaF crystals

Soft glass capillaries cover the complete range of all polarities of the stationary phases. The main significances are to
be found when using soft glass capillaries in lengths between
50 and 100 meters. By taking care of the upper temperature
limits, they may have a longer durability. HCl or HCl/H_2F_2-
treatment is not possible with borosilicate glass as not enough
crystals are produced.

The most effective deactivation of borosilicate glass is done
by persilylation as described by Grob. The disadvantage of this
method is the low yield. Before persilylating, leaching has to
be carried out. Just methylpolysiloxanes are coatable. Otherwise, instead of 1,3-diphenyl-1,1,3,3-tetramethyldisilazane
1,3-dimethyl-1,1,3,3-tetraphenyldisilazane or similar silylating
reagents have to be used (3,14,19). Deactivation by cyclic siloxanes is possible. It is important that these cyclic siloxanes
are available as fluoro-, chloro-, or cyano-derivatives, which
means that a coating with polar polysiloxanes is possible by

using these polysiloxanes as deactivation reagents.

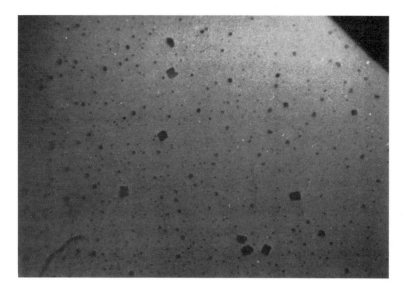

Fig. 14 HCl-treatment of borosilicate glass

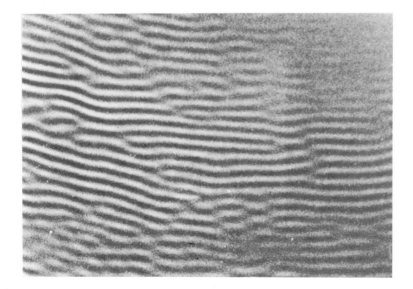

Fig. 15 Silylated borosilicate glass

The separation power of borosilicate glass capillaries is identical to soft glass and fused silica capillaries. The inactivity is better than with soft glass, and similar to fused silica. A wide range of stationary liquid phases are coatable, but not so many as with soft glass and even more than with fused silica. The upper temperature limit is the highest one of all materials discussed here. From this point of view, the borosilicate glass capillaries are suitable for High-Temperature-Chromatography above +350°C. The durability is similar to soft glass and longer than for fused silica.

The deactivation of fused silica capillaries should be done by cyclic siloxanes. The separation power is identical with borosilicate glass and soft glass capillaries. The main advantage of fused silica capillaries is their flexibility. We would like to stress the point once again that they may fracture without being noticed by the user due to the outer varnishing.

The basis of the comparison of these three column wall materials is the constancy of all parameters. In all cases of comparison the same coating procedure, the same column wall diameter, and the same injection system were used. Generally, it can be stated: the separation power of the so-called High-Resolution-Capillaries is similar to the separation power of those capillaries which have been used in Europe for nearly 15 years. During the last years only the inertness has increased.

Permanent bonding of stationary phases is possible with all column wall materials discussed here. For crosslinking, γ-radiation is very useful but not practicable for everybody. Within the γ-radiation the column wall materials vary; this can be seen by a loss of absorption beyond 300 nm and intense absorption band at 215 nm. Sometimes, dark violet coloured glass walls can result. Simultaneously, a loss of mechanical resistance can be observed.

For crosslinking we recommend the treatment with aromatic
peroxides (catalytic effect). We generally use stationary
phases for crosslinking with an amount of at least 4% vinyl
groups. The amount of aromatic peroxide should be in the range
of about 1% of the amount of stationary phase.

Crosslinking of non-vinylated polysiloxanes is possible only
by using the highest amounts of peroxides. The molecular weights
of such polysiloxanes should be in the range of $6 \cdot 10^6$. For the
crosslinking of phenyl polysiloxanes it is essential to have
vinyl groups, otherwise the linking will not be effective.

Our crosslinked stationary phases are coupled with the column
wall as we use the chloro-vinyl-silazane for bonding. This
bonding can be observed by extinguishing the band at 3750 cm^{-1}
(vibrating stretch band of isolated OH). Permanent bonded
phases are the most useful for on-column injection. The separation power in the lower temperature range is a little less
than that of normal coated stationary phases. The main argu-
ment for using permanent bonded phases is the impossibility
of thermal rearrangement, i.e. no loss of separation power.
Most important is the possibility of re-employing solvents for
cleaning columns when using on-column injection. In our crosslinking procedure, no deactivation is necessary. Moreover
we found out that the most effective pretreatments for deactivation are(10,11): 1. when using soft glass Carbowax 20 M-
treatment (12), 2. borosilicate glass, persilylation after
static leaching, 3. fused silica, deactivating with cyclic
siloxanes. It can easily be seen that the crosslinking runs
over Si-C-C-Si-bondings.

Generally, soft glass is coatable with every stationary phase,
borosilicate glass nearly with all stationary phases. Better
results are yielded with high viscosity phases, especially
the gum phases, such as OV-1, SE-30, SE-53, SE-54, and especially the polar phases Carbowax 20 M, WG-11, OV-351 and simi-

lar phases.

Fused silica should only be coated with gum phases, OV-1701, WG-11, OV-351, or similar phases. Other low viscosity polysi- loxanes should not be used due to their too low durability.

In comparison: a Carbowax 20 M column coated on soft glass or borosilicate glass has a durability of one year; coated on fused silica one can estimate at least 3-4 weaks in daily use, e.g. 15 analyses a day of ethylvanilline in a fragrance. The highest upper temperature limit is found with OV-1 and SE-52. Borosilicate glass is good for + 400 up to 475°C, soft glass up to + 220°C, sometimes 250°C, fused silica up to + 300°C, sometimes 350°C. The bleeding rate of fused silica is normally higher than that one of borosilicate. The next two figures 16 and 17 show the difference between crosslinking and chemical bonding to the column wall surfaces. Fig. 16 shows the MS of just crosslinked OV-1 on Pyrex at + 400°C; Fig. 17 shows the same stationary phase at 450°C, but chemically bonded to the Pyrex surface. Beyond these temperatures no MS underground is visible.

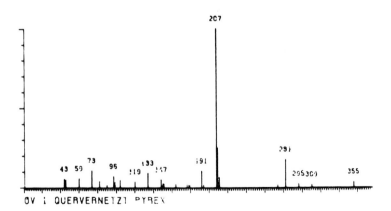

Fig. 16   OV-1 crosslinked without chemical bonding to the
          glass surface

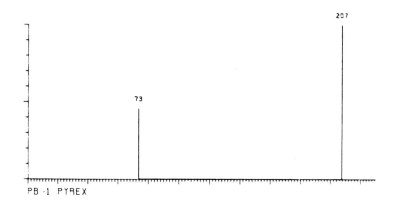

PB -1 PYREX

Fig. 17   OV-1 crosslinked and bonded to the glass surface

The next figure gives a comparison of the influence of the
column wall pretreatments with 2 selected stationary phases.
We are still working on the exact explanation of the shown
index differences.

	Indices									
	WG-11					OV-101*				
	X	Y	U	Z	S	X	Y	U	Z	S
Soda lime glass	961	929	914	1159	1207	650	491	579	531	731
Pyrex	971	923	917	1167	1207	674	427	580	534	741
Duran	972	931	920	1170	1211	659	438	574	543	727
Fused silica	986	943	900	1169	1206					
						*Silanized				

Fig. 18   Influence of column wall pretreatments on index
          differences

Conclusion: The simplest deactivation is possible by using fused silica. The coating of fused silica capillaries is very easy because of their thin walls (temperaturing). Also the fused silica columns can be handled in the easiest way because of their flexibility. This is the reason why they are mostly offered, but it does not mean that they are the most useful ones.

References

1. Günther, W., Schlegelmilch, F., Hecker, K., Klockenkämper, H.: Labo 8 (1979).

2. Rohrschneider, L.: J. Chromatogr. 22, 6 (1966).

3. Watanabe, C., Tomita, H.: J. Chromatogr. Sci. 13, 123 (1975).

4. Grob, K., Grob, G.: J. Chromatogr. 125, 471 (1976).

5. Alexander, G., Garzo, G., Palyi, G.: J. Chromatogr. 91, 25 (1974).

6. Tesarik, K., Novotny, M., Struppe, H.G. (Eds.): Gas-Chromatographie 1968, Akademie-Verlag GmbH, Berlin, 575, 1968.

7. Alexander, G., Rutten, G.A.F.M.: Chromatographia 6, 231 (1973).

8. Novotny, M., Tesarik, K.: Chromatographia 1, 332 (1968).

9. Novotny, M., Bartle, K.D., Blomberg, L.: J. Chromatogr. Sci. 8, 390 (1970).

10. Rutten, G.A.F.M., Luyten, J.A.: J. Chromatogr. 74, 177 (1972).

11. Badings, H.T., Van der Pol, J.J.G., Wassink, J.G.: Chromatographia 8, 440 (1975).

12. Schomburg, G., Husmann, H., Weeke, F.: J. Chromatogr. 99, 63 (1974).

13. Grob, K., Grob, G., Grob jr., K.: Chromatographia 10, 181 (1977).

14. Grob, K., Grob, G., Grob jr., K.: HRC & CC (2), 677 (1979).

15. Roukeria, S.: Diplom-Arbeit FH Niederrhein, Krefeld, 1983.

THE USE OF ATD-50 SYSTEM WITH FUSED SILICA CAPILLARIES IN DYNAMIC HEADSPACE ANALYSIS

Jan Kristensson

Department of Analytical Chemistry, University of Stockholm, S-106 91 Stockholm, Sweden

Introduction

The use of passive sampling and thermal desorption was adopted in Great Britain by a number of organisations who collectively set up the Personal Monitor Users Committee (Working Group 5 of the Health & Safety Executive Committee on Analytical Requirements). This committee specified the parameters for the diffusive sampling system which is the basis for the Perkin-Elmer ATD-50 system (1). The ATD-50 system combines passive or pumped sampling with fully automatic thermal desorption and gas chromatographic analysis. The system was mainly developed for occupational hygiene measurements of volatile organic compounds, but we have also used it for measurements of volatiles in ambient air. Both applications will be discussed below.

The first system for sampling by diffusion was described by Palmes and Gunnison in 1973 (2). Since then several diffusion sampling systems have been developed, and different samplers based on diffusion and permeation are reviewed by Rose and Perkins (3).

Sampling

Diffusive sampling is based on Fick's first law of diffusion,

$$N = -DA \frac{dc}{dx} \qquad (1)$$

where

    N = diffusive transport rate, moles/sec

    D = diffusion coefficient in air, cm²/sec

    A = diffusion path cross-section area, cm²

    c = concentration, moles/cm³

    x = distance, cm .

Under steady state conditions and integrating equation 1 over the length of the diffusion zone, L (cm), the amount of pollutant collected, M (moles), during the time t (sec) is described by

$$M = \frac{DA}{L} \, Ct \, . \qquad (2)$$

The following assumptions are made:
1. concentration at adsorbent surface is zero (zero sink);
2. concentration at sampler face represents the ambient concentration;
3. diffusion coefficient is independent of concentration and matrix effects;
4. steady state diffusion applies.

The ATD-50 sampler consists of a stainless steel tube (90 mm x 5 mm ID) (Figure 1); a stainless steel gauze in one end defines the diffusion zone length to 15 mm. The sampler is packed with a suitable adsorbent, e.g. Tenax, Chromosorb or Amberlite which is fixed in the tube by a gauze and a spring.

For diffusion sampling the storage cap at the end with the diffusion zone is changed by a diffusion cap, e.g. a cap with a stainless steel gauze. For pumped sampling both storage caps are removed and the tube is connected to a sampling pump. A diffusion cap can be used to protect the particles to enter the sampling tube. The main advantage with diffusion sampling is that pumps are not needed. Thus more samples can be collected easier and cheaper than by other systems.

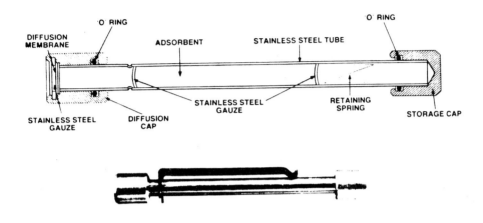

Figure 1 ATD-50 sampling tube

ATD-50 system

The ATD-50 system is designed for automatic thermal desorption of up to 50 samples a time. Before desorption the storage caps are changed by caps with ball valves, and the tubes are placed on a turn table. The thermal desorption can be carried out directly from the adsorbent tube into the analytical column (single stage) or via a cold trap (two stages). In the latter case, the sample is injected as a narrow band into the analytical column by rapid heating of the cold trap (approx. 20°C/sec). The cold trap is cooled by a peltier cooler. The parameters to be set at the automatic thermal desorption are shown in Table I.

Before thermal desorption the air in the tube is flushed out with nitrogen, and the tube is tested for leaks against the pre-

set pressure. If the tube fails in the test a fail signal is
given and the tube is not desorbed. If the tube passes the leak
test it is desorbed at a preset temperature and time into the
cold trap. The cold trap is heated and the sample is transferred via the heated transfer line (1/16 inch stainless steel
tube) into the gas chromatograph for analysis. Any gas chromatograph can be connected to the ATD system.

Table I ATD-50 parameters

Parameter		Maximum	Minimum
Method		1	4
Mode	(Desorp. modes)	1	7
Oven	(Desorp. temp.,°C)	50	250
Desorb	(Desorp. time,min)	3	30
Box	(Transfer line temp. °C)	50	150
CTL	(Cold trap low, temp. °C)	-30	30
CTH	(Cold trap high, temp. °C)	50	300
Anal	(Analysis time,min)	4	99
First	(First tube ,nr.)	1	50
Last	(Last tube ,nr.)	1	50
Press	(Inlet pressure ,psi)	-	-
Cycle	(Total cycle time,min)	-	-

Modification of the ATD system

In most applications with the ATD packed columns have been used.
To achieve more efficient separations it is desirable to use
capillary columns. The long heated transfer line has a big dead
volume, and the influence of that dead volume can be minimized
by a split system after the transfer line, but still adsorption risks in the stainless steel tube exist. To overcome
these risks our ATD system was modified as follows: A fused silica interface (0.8 m x 0.22 mm) deactivated with cyclic siloxanes (4) was inserted into the transfer line and into the cold

trap (Figure 2). At the end of the transfer line a split valve
was connected by a union for fused silica columns (Valco Instruments Co. Inc.).

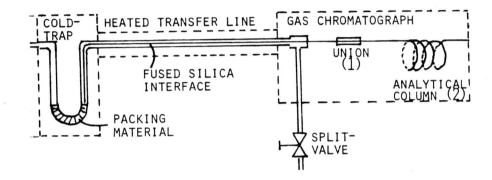

Figure 2 Connection of fused silica interface and column to
         ATD-50
         1. Union for fused silica (Valco Instruments Co.)
         2. Fused silica analytical column

Effects of modification

This modification facilitates the connection of any capillary
column to the system; it eliminates the peak broadening and adsorption effects in the transfer line and allows to adjust injected volumes by changing the split ratio (no problems with
inhomogenous splitting exist). Thus, it is possible to recol-
lect the split effluent and to use a second injection of the
same sample. Consequently, the main disadvantage with thermal
desorption is avoided (5). No peak broadening and no adsorption effects are remarked in the modified system (Table IIa
and IIb; Figures 3a-c).

114

3a. split ratio 5:1

3b. split ratio 10:1

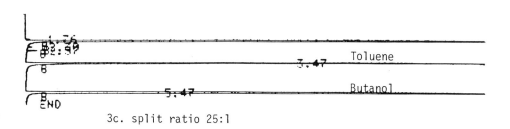

3c. split ratio 25:1

Figure 3 Examples of chromatograms at different split ratios
a = 5:1, b = 10:1 and c = 25:1
Column: CP Wax 57 CB, 25 m x 0.32 mm, 1 μm

Table II

Effects of peak broadening and adsorption at different split
ratios. 172 ng toluene and 162 ng butanol were injected in
each sample independent of the split ratio. The relative peak
width was calculated from the values from the integrator.
$w_{0.5}$ = area/height, arbitary units.

Table IIa Fused Silica column, SE-30, 25 m x 0.22 mm

Split-ratio	Compound	Retention time (min) I	II	Rel. peak width I	II
1/5	Toluene	3.53	3.54	2,71	2,58
	Butanol	5.85	5.79	3,68	3,80
1/10	Toluene	3.54	3.52	2,97	2,45
	Butanol	5.84	5.73	3,74	3,79
1/25	Toluene	3.58	3.61	2,50	2,53
	Butanol	5.73	5.76	3,80	3,97
1/50	Toluene	3.97	3.74	2,85	2,80
	Butanol	5.89	6.18	3,67	4,15

Table IIb Fused Silica column, CP Wax 57, 25 m x 0.32 mm

Split-ratio	Compound	Retention time (min) I	II	Rel. peak width I	II
1/5	Toluene	3.42	3.42	2,01	1,95
	Butanol	5.45	5.51	4,58	4,42
1/10	Toluene	3.43	3.43	1,97	2,08
	Butanol	5.55	5.55	4,56	4,52
1/25	Toluene	3.47	3.46	2,83	2,52
	Butanol	5.47	5.45	4,10	4,10
1/50	Toluene	3.58	3.65	2,73	2,94
	Butanol	5.60	5.79	3,88	4,00

Results

Applications where ATD has been used with capillary columns can
be found in four application notes from Perkin-Elmer:
Analysis of petrol vapour with packed SCOT and WCOT columns
(a comparison) (6); analysis of power station stack effluents
(7); analysis of ethylene glycol in north sea gas (Mono ethylene glycol is added as a plasticiser to the north sea gas,
preventing pipeline seals from drying and fracturing) (8); analysis of residual solvents in adhesive tapes (9). Three applications where we have employed capillary columns with the ATD
system will be discussed here:
A. analysis of solvent vapours in a printing industry
B. analysis of aromatic hydrocarbons in the environment
C. analysis of anesthetics in operating theaters.

A. Analysis of solvent vapours in a printing industry

The task of this investigation was to study the volatiles in
the ambient air in a printing industry where workers were exposed to solvent vapours. The study was the basis for subsequent blood platelet examinations (10).

ANALYTICAL CONDITIONS:

ATD-50		Sigma 2 B	
Desorb. temp:	250°C	Column:	25 m x 0.32 mm FS
Desorb. time:	3 min		CP Wax 57 bp, 1.2μm
Cold trap low:	-30°C		1.8 ml/min, split 50:1
Cold trap high:	250°C	Temp:	50°C 2 min, 4°C/min
Cold trap packing:	Tenax		to 100°C, 8°C/min to
Inlet pressure:	20 psi		200°C
		Detector:	FID at 250°C

An example of a typical chromatogram is given in Figure 4. Calculations for toluene, xylene isomers and trimethylbenzene isomers gave approximate concentrations from 0.1 to 20 ppm. These
compounds represented approx. 50% of the total concentration.

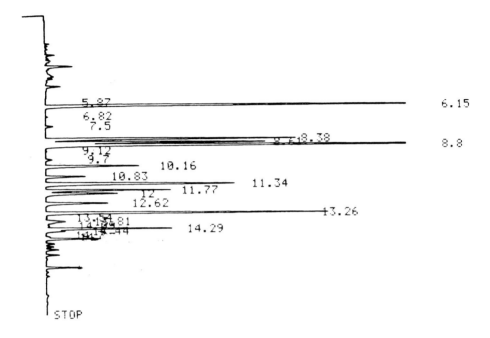

Figure 4   Example of a chromatogram from a diffusive sample
          taken at a printing industry

B. Analysis of aromatic hydrocarbons in environment

In this investigation diffusive sampling was compared with
pumped samples from a suburb of Stockholm. The sampling place
was considered exhibiting "clean air" (reference place for
measurements inside the center of Stockholm).

ANALYTICAL CONDITIONS:

ATD-50		Sigma 2B	
Desorb. temp:	250°C	Column:	25 m x 0.32 mm FS
Desorb. time:	3 min		CP Wax 57 bp, 1.2μm
Cold trap low:	-30°C		1.8 ml/min, split 5:1
Cold trap high:	250°C	Temp:	50°C 2 min, 4°C/min
Cold trap packing:	Tenax		to 100°C, 8°C/min to
Inlet pressure:	20 psi		200°C
		Detector:	FID at 250°C

Pumped ATD tubes for 5.5 hours (5.3 l) and 1 hour (2.5 l),
cp. Fig.5a-b, were compared with a diffusive sample for 8 hours,
Figure 5c. The concentration was approx. 0.3 ppb for toluene.
The samples showed that is was possible to take diffusive samples even at very low concentrations.

C. Analysis of anesthetics in operating theaters

To study the local concentration of anesthetics in operating
theaters nitrous oxide was analysed after diffusive sampling.
The adsorbent in the ATD tube in this case was molecular sieve
13X, 60/80 mesh. To separate nitrous oxide from oxygen and CO_2
it is necessary to use a GSC column. A micropacked column
(1.6 m x 1.6 mm OD, 0.7 mm ID) packed with Porapak QS 100/120
mesh was connected to the fused silica interface. Nitrous
oxide was detected with an electron capture detector.

ANALYTICAL CONDITIONS:

ATD-50		Sigma 2 B	
Desorb. temp:	150°C	Column:	1.6 m x 0.7 mm micro-
Desorb. time:	3 min		packed, Porapak QS,
Cold trap low:	-30°C		100/120 mesh
Cold trap high:	200°C		1.8 ml/min, split 50:1
Cold trap packing:	Mol.Sieve	Temp:	80°C isothermal
Inlet pressure:	10 psi	Detector:	ECD at 350°C

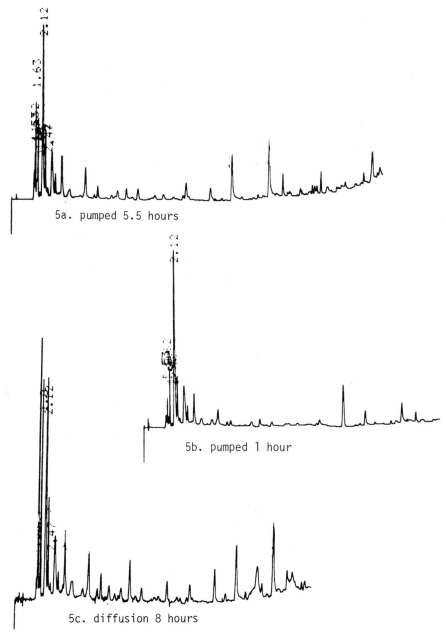

5a. pumped 5.5 hours

5b. pumped 1 hour

5c. diffusion 8 hours

Figure 5   Chromatograms from ATD samples from ambient air
           5a   Pumped ATD tube 5.5 hours (5.3 l)
           5b   Pumped ATD tube 1 hour (2.5 l)
           5c   Diffusive ATD tube 8 hours

With the described system it was possible to analyse nitrous
oxide in the concentration range 1 to 200 ppm using eight
hours diffusion sampling. The ATD system for nitrous oxide has
been evaluated together with Siemens Nitrous Oxide Monitor (11).
An example of a chromatogram is shown in Figure 6.

Figure 6   Example of a chromatogram from diffusive sampling of
           nitrous oxide from an operating theatre, 98 ppm

References

1. Perkin-Elmer: Model ATD 50 Thermal Desorption Application
    No. 1.

2. Palmes, E.D., Gunnison, A.F.: Am. Ind. Hyg. Assoc. J. 34,
    78 (1973).

3. Rose, V.E., Perkins, J.L.: Am. Ind. Hyg. Assoc. J. 43, 605
    (1982).

4. Blomberg, L., Buijten, J., Markides, K., Wännman, T.: HRC
    & CC 4, 578 (1981).

5. Kristensson, J., Sjölund, N.: in prep.

6. Perkin-Elmer: Model ATD 50 Thermal Desorption Application
    No. 2.

7. Perkin-Elmer: Model ATD 50 Thermal Desorption Application
    No. 6.

8. Perkin-Elmer: Model ATD 50 Thermal Desorption Application
    No. 7.

9. Perkin-Elmer: Model ATD 50 Thermal Desorption Application
    No. 8.

10. Beving, H., Malmgren, R., Olsson, P., Tornling, G., Unge,
    G.: Scand. J. Work Environ.Health in press.

11. Kristensson, J., Sjölund, N.: Research report. The Swedish
    Work Environment Found. Contract No. 82-1156, in press.

COUPLED GAS CHROMATOGRAPHIC METHODS FOR SEPARATION, IDENTIFICATION, AND QUANTITATIVE ANALYSIS OF COMPLEX MIXTURES: MDGC, GC-MS, GC-IR, LC-GC

G. Schomburg, H. Husmann, L. Podmaniczky and F. Weeke

Max-Planck-Institut für Kohlenforschung and
Max-Planck-Institut für Strahlenchemie,
4330 Mülheim-Ruhr, FRG

and in cooperation with

A. Rapp

Bundesforschungsanstalt für Rebenzüchtung Geilweilerhof
6741 Siebeldingen über Landau/Pfalz, FRG

INTRODUCTION

In flavor and fragrance analysis mixtures arise that may consist of very many species with wide range of polarities as well as molecular sizes (carbon numbers). The interesting and significant groups of components may occur in various homogenous or heterogenous liquid or solid matrices from where they must be transferred into a new matrix since most of the components of such matrices are not compatible with GC. Major constituents of the matrix, such as water, may also disturb the gas chromatographic separation, even when they are volatile enough. Without preseparational procedures such as extraction, vaporization, (head-space GC), LC, etc. the obtained samples usually contain, for instance, non-volatiles or other disturbing matrix components. Certain species that are of importance for the character or the intensity of an aroma may be present in very low concentrations and moreover, may be analytically masked, i. e., chromatographically overlapped by other components of little or none importance regarding the aroma. When using the common GC detectors in flavor analysis, it may also be difficult to detect constituents of high olfactoric intensity or to obtain a mass infrared or NMR-spectrum of such compounds. To achieve this, enrichment of these compounds has to be done. Aromas may have to be either qualitatively and/or quantitatively analyzed. One or several unknowns may have to be isolated and identified when they are expected to be of significance as a flavor component. A total "pattern" analysis for both, the known and unknown components, may be of great interest in order to reveal, for example, falsification or generally for the characterization of commercial products. This step has to be carried out by the listing of all the retention data and the peak areas (uncorrected or corrected by response factors) of the constituents and of the standard compounds used. These data should have been measured at sufficient or adequate precision and accuracy. Various basic difficulties may arise with the separation, the identification, and the quantitation of such complex mixtures, when only applying standard GC instrumentation consisting of a single sampling device, a single column and a single detector:

- The resolution (separation efficiency and selectivity) as well as the peak capacity may not be sufficient enough regarding the large number of components, especially in the case of extreme concentration ratios of neighboring chromatographic species.

- The preparative isolation of certain unknown species may be impossible to achieve in a single run because of too low column capacity if the high resolution of capillary columns is required, much larger amounts of substances are needed for NMR and IR identification, inspite of extended use of the Fourier-technique; even with GC-MS low concentration components may not be detectable, especially not those that have a high olfactoric intensity.

- Retention data for characterization of the composition of mixtures cannot be measured reproducibly enough for Table and literature search applications.

- The quantitative (peak area) data suffer from systematical errors of the GC analysis due to inadequate sampling techniques, column technology, or detection.

It is going to be discussed what kind of instrumental and methodical progress has recently been established in capillary GC concerning the above mentioned limitations, particularly on the basis of achievements carried out in the authors own laboratories.

Because of their complexity, aroma samples can only be analyzed and evaluated if the optimum of information regarding the qualitative and quantitative composition and also the necessary information concerning the chemical structure of selected significant species (including unknowns) is achievable. Nowadays, chromatographic (retention and response) information can be obtained with improved reproducibility when applying modern equipment and column technology. Such chromatographic information on component identities may be valuable, although spectroscopic information originating from MS, NMR, IR, etc. regarding the single separated species is absolutely necessary, especially to safely elucidate the structures of unknowns. In most cases the low concentrations of such constituents of interest and the complexity of the matrices give cause to difficult chromatographic separations. In such cases the required spectroscopic information is complicated to achieve for several reasons, and retention data may then be very useful, too.

INSTRUMENTATION FOR STANDARD AND MULTIDIMENSIONAL GC AT HIGH RESOLUTION AND REPRODUCIBILITY

The chromatographic systems of commercially available equipment have been considerably improved for high resolution work in capillary columns in the recent years, also regarding sampling, column technology and column connection, as well as detectors.

Sampling techniques:

Cold sample introduction into the column directly[1-3] or into a temperature programmable external vaporizer[4-6] are to be preferred for several reasons which are:
- gentle conditions during the vaporization
- improvement of precision and accuracy, i.e., avoidance of

discrimination such as changes of the original compositions
- improvement of resolution in the neighborhood of the solvent peak (solvent effects)[7].

Essential oil or flavor samples are the most favorable sample types for any kind of "splitless" sampling techniques since they are in most cases already diluted or contain so many components that further intensive dilution for reasons of sampling becomes unnecessary.

Column technology:

Major interesting points in recent achievements in this area are the manufacture of columns with a reproducible performance regarding efficiency, deactivation of support surface, and especially regarding polarity. By crosslinking and immobilization of various types of non-polar and polar polysiloxanes[11-15] as well as of high molecular weight polyethylene glykols (Carbowax 20 M)[16-17] columns with more homogeneous and stable films are obtained that offer in practical work the following advantages:
- high efficiency of columns with < 1 μ coatings
- high sample capacity, efficiency and film stability in particular of thick film (>1 μ) columns that can preferably be produced by using the non-polar polysiloxanes and the Carbowax 20 M
- avoidance of deterioration of the coating caused by droplet formation and displacement of the stationary liquid in the case of sampling of large volumes of polar solvents or major components, especially when using Carbowax 20 M columns
- stable polar polysiloxane columns of adequate efficiency containing mainly CN-groups.

In Fig. 1, 2, and 3 gas chromatograms of wine aroma extracts (origin Prof. A. Rapp, Geilweilerhof), the corresponding artificial mixtures of a selection of important wine aroma components and of a beer wort extract (origin Dr. Nitz, Weihenstephan) illustrate advantages and disadvantages of thick film non-polar alkylpolysiloxane columns in their application to essential oil analysis. From Fig. 3 it can be concluded that further

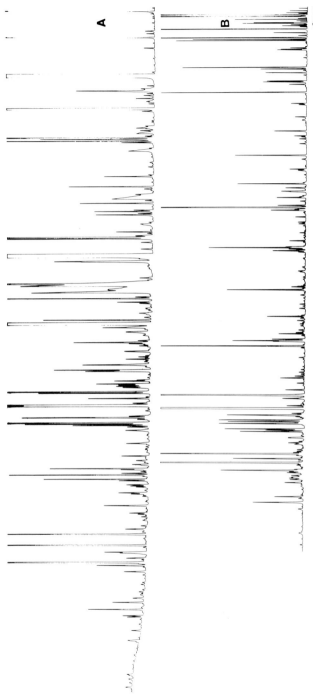

SEPARATION OF WINE AROMA EXTRACT (A) AND BEER WORT EXTRACT (B) BY USING A NON-POLAR

THICK FILM METHYLPOLYSILOXANE (OV 1) COLUMN.

Peak symmetry of components of very different polarities.

Sample   : **A** Wine aroma extract, origin A. Rapp   **B** Beer wort extract, origin Dr. Nitz

Column   : 50 m methylpolysiloxane OV 1, 0.27 mm i. d., $d_f = 1.5$ µm

Temperature : **A** 30–280 °C, 3 °C/min   **B** 30–250 °C, 3 °C/min

Carrier gas : **A** 0.7 bar H_2 **B** 0.8 bar H_2 Instrument : Double oven, Siemens Sichromat 2

G. Schomburg, H. Husmann, L. Podmaniczky, F. Weeke, Max-Planck-Institut für Kohlenforschung, August 1983

Fig. 1

126

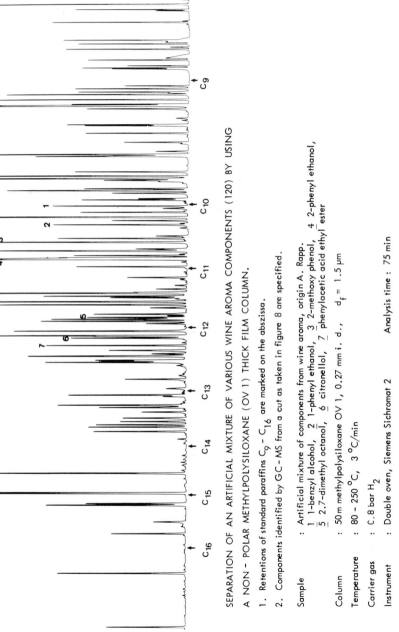

SEPARATION OF AN ARTIFICIAL MIXTURE OF VARIOUS WINE AROMA COMPONENTS (120) BY USING
A NON - POLAR METHYLPOLYSILOXANE (OV 1) THICK FILM COLUMN.

1. Retentions of standard paraffins C₉ - C₁₆ are marked on the abszissa.

2. Components identified by GC-MS from a cut as taken in figure 8 are specified.

Sample : Artificial mixture of components from wine aroma, origin A. Rapp.
1 1-benzyl alcohol, 2 1-phenyl ethanol, 3 2-methoxy phenol, 4 2-phenyl ethanol,
5 2.7-dimethyl octanol, 6 citronellol, 7 phenylacetic acid ethyl ester

Column : 50 m methylpolysiloxane OV 1, 0.27 mm i. d., d_f = 1.5 μm

Temperature : 80 - 250 °C, 3 °C/min

Carrier gas : C. 8 bar H₂

Instrument : Double oven, Siemens Sichromat 2   Analysis time : 75 min

G. Schomburg, H. Husmann, L. Podmaniczky, F. Weeke, Max-Planck-Institut für Kohlenforschung, August 1983
Fig. 2

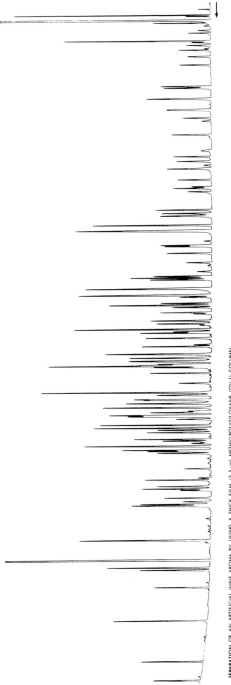

SEPARATION OF AN ARTIFICIAL WINE AROMA BY USING A THICK FILM (3.5 μm) METHYLPOLYSILOXANE (OV 1) COLUMN.

Sample : 0.5 μl wine aroma   Split : 1:40

Column : 55 m PS 240 (methylpolysiloxane), 0.27 mm i. d., d_f = 3.5 μm

Temperature : 30 - 250 °C, 3 °C/min

Carrier gas : 1.0 bar H_2 Analysis time : 85 min

G. Schomburg, H. Husmann, L. Podmaniczky, F. Weeke, Max-Planck-Institut für Kohlenforschung, August 1983

Fig. 3

increase of the stationary liquid content improves the resolution of the more volatile constituents, etc. but a serious decrease of separation efficiency can also be recognized from the resolution of certain peak multipletts. The major disadvantage of this weakly polar column type seems to be the overloading by the highly polar solutes, especially free acids, as contained in wine aroma extracts which leads to asymmetrical peaks with "leading". This fact is demonstrated on hand of the chromatograms of wine aroma extract as obtained by using a non- (or weakly) polar thick film methylpolysiloxane column, or for comparison using other, polar stationary phases such as: Ucon, Carbowax 20 M, and Reoplex 400.

A specific problem involved in the separation of wine aroma extracts are large concentrations of alcohols, diols, as well as free acids. Using the non-polar polysiloxane columns, the deactivation of the support surface of the highly polar acids can be easily carried out, but the limited sample capacity of these phases causes overloaded peak shapes or prevents the detection of other low concentration components if excessive overloading has to be avoided. In Fig. 4 the double trace chromatograms of a multidimensional separation of a wine aroma extract are given. The upper chromatogram was obtained using a 25 m polar Ucon column, whereas the chromatogram located below originates from the cut taken from the pre-separation as marked and transferred into the non-polar OV-column. The Ucon chromatogram contains peaks with tailing (free acids, alcohols, etc.). In the OV 1 chromatogram, an excellent resolution of the compounds contained in the relatively narrow cut is observed. The free acid which was also present in the cut exhibits a strong leading effect because of poor solubility in the non-polar stationary phase.

In using the polar stationary liquids, the deactivation (persilylation [18-19], PSD [20-21], D_4 [22], etc.) methods which are effective with the production of good polysiloxane columns cannot be applied. For the improved separation of the free acids an acidification of the surface can be done by additives. Immobilization of polar stationary phases which is the favored method for achieving stable and homogenous films could, also be obtained by crosslinking of polyethylene glycols Carbowax 20 M. The polar cyanopolysiloxanes may also be

129

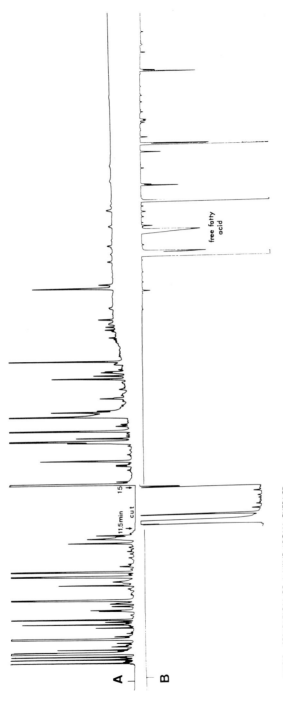

MDGC - SEPARATION OF A WINE AROMA EXTRACT

Polar (Ucon) pre- and non-polar (methylpolysiloxane OV 1, thick film) main-column.

Sample   : 0.3 ul wine aroma extract

Column   : **A** 25 m Ucon 75 H 90000   **B** 50 m methylpolysiloxane OV 1, d_f = 1 - 1.5 um

Temperatures : **A** 80 °C, to 240 °C, 6 °C/min   **B** 0 °C, 13 min iso, to 60 °C, 15 °C/min, to 250 °C, 6 °C/min

Carrier gas : H_2 Analysis time : 50 min   Instrument: Double oven, Siemens Sichromat 2

G. Schomburg, H. Husmann, L. Podmaniczky, F. Weeke, Max-Planck-Institut für Kohlenforschung, August 1983

Fig. 4

immobilizable by crosslinking, chemical bonding or in situ oligomerization[23]. The flexible fused silica columns[24] have facilitated the instrumental problem of column manipulation and connection; the GC-MS interfacing also has become technically much simpler through direct introduction of the column outlet into the ion source[25]. Moreover, fused silica tubing is generally very suitable for any kind of inert and easy to establish carrier gas flow lines. The material is not as flexible, however, that coiling diameters of column of less than lets-say 120 mm are not applicable in permanent use, inspite of the polyimid coating of the material. The thin walled fused silica columns moreover suffer from the disadvantage of fast heat transfer from the column oven onto the stationary and mobile phase of the column. Especially local temperature fluctuations occuring in most of the common commercially available column ovens, lead to peak splitting and broadening[26] (christmas tree effect)[16]. Therefore new, in this respect better performance column ovens are needed in the future. The heat capacity of the column itself can also be increased by special measures[26].

DESIGN OF GC INSTRUMENTATION IN GENERAL, COLUMN OVENS, FLOW-SYSTEMS, COLUMN COUPLING AND CONNECTION, etc.

Proper temperature control and adjustment, not only of the column alone but also within the different important regions of the chromatographic system are the most important parameters to be optimized for difficult separations and quantitative analyses. Concerning the column, temperature programming during the separation is the most important approach of selectivity adjustment for the large number of different solute pairs in complex mixtures. Temperature programming has also proved to be of advantage when applied to external vaporization during sampling[4-6]. Multi-column oven instruments open up new possibilities of optimization for analysis time, resolution, and the detection limits of traces, in case systems of coupled columns are used[27]. Besides from the typical multidimensional applications a second, independently operable but integrated column oven can also be applied for reaction GC, or for pre-column sampling[28]. The capabilities of such MDGC systems are only optimally used, when each column is operated in separate ovens, either isothermally and/or temperature programmed. The coupled columns may differ with regard to either stationary phase content (film thickness), polarity, and efficiency (length, diameter). By transfer of selected cuts from the first into the second column, strong changes of capacity ratios, selectivities and efficiencies can be achieved[29]. Advantage of these achievements can only be taken, if a partial analysis of broad or narrow cuts that yields the needed significant information for the characterization of flavors, is carried out. Also analysis time can thereby be saved.

OPTIMIZATION AND APPLICATION OF MULTIDIMENSIONAL SYSTEMS: ANALYSIS OF WINE AROMA

A practical aim of analytical efforts in essential oil analysis, besides from the discovery and identification of overlapped and/or low concentration components is the characterization of mixtures by retention and response data, obtained for selected sections of the entire chromatographic pattern. With different polarities of the coupled columns, multiple data sets can be generated, provided the resolution achieved with the pre-separation is good enough and whether retention measurements with each column can be done independently. By using a polar pre-column, a complete resolution of all the components must not necessarily be obtained thereby, a group-type separation according to solute polarity may be sufficient. Overlappings of peaks are removed by the change of selectivity, efficiency, and phase ratio of the second column, if also the range of the transferred cuts (at the retention axis of the chromatogram) is adapted to the number of constituents prospectively contained in this region of the pre-separation chromatogram.

A group of certain species of similar retentions contained in a cut of limited broadness is transferred into a column of different polarity whereby the greatest change of polarity is, of course, achieved when the second column is a non-polar polysiloxane one (see Fig. 4 and 5). From the chromatograms of Fig. 5 it can be recognized that by polarity change

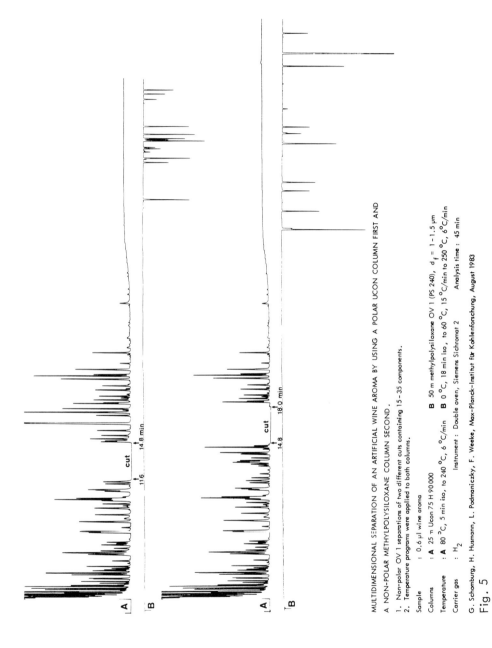

MULTIDIMENSIONAL SEPARATION OF AN ARTIFICIAL WINE AROMA BY USING A POLAR UCON COLUMN FIRST AND

A NON-POLAR METHYLPOLYSILOXANE COLUMN SECOND .

1. Non-polar OV 1 separations of two different cuts containing 15 - 35 components.
2. Temperature programs were applied to both columns.

Sample   : 0.6 µl wine aroma

Columns   : **A** 25 m Ucon 75 H 90 000   **B** 50 m methylpolysiloxane OV 1 (PS 240), d_f = 1 - 1.5 µm

Temperature   : **A** 80 °C, 5 min iso, to 240 °C, 6 °C/min   **B** 0 °C, 18 min iso, to 60 °C, 15 °C/min to 250 °C, 6 °C/min

Carrier gas   : H$_2$ Instrument : Double oven, Siemens Sichromat 2   Analysis time : 45 min

G. Schomburg, H. Husmann, L. Podmaniczky, F. Weeke, Max-Planck-Institut für Kohlenforschung, August 1983

Fig. 5

a tremendous increase in resolution can be achieved. The shapes of the separated peak in the non-polar gas chromatogram are excellent. Consequently many more constituents can be separated and identified by adopting such a multidimensional separation procedure. Such a system was investigated with regard to its practical usefulness in essential oil analysis. If the second column is exchanged with a column of different polarity, another set of different retention data can be obtained. Using this procedure retention data of high reproducibility are only determined within the second column.

In this work, experiments on the following approaches and problems were carried out:

- Temperature programming with both columns, the usage of retention data originating from temperature programmed separations that disadvantageously depend on more parameters than the isothermal Kovats indices such as carrier gas flows initial temperatures, program-rate, and film thickness of the column.

- Trapping between the first and the second column in order to avoid mixed retentions, the sectional trapping within the inlet of the second column, the trapping of very volatile components,

- Retention index measurements within the second column, the necessity of perfect trapping of the volatile sample constituents and the suitable paraffin standards for the index calculation.

The trapping within the second column becomes relatively easy to handle when temperature programming is to be applied, which involves low and even subambient initial column temperatures. With the Sichromat 2, the second oven can easily be cooled down automatically by using liquid nitrogen. The column inlet of the second column can, of course, be additionally cooled down by the common CO_2 or the liquid N_2 trapping devices[30], as had been described in one of our own papers some years ago. Such trapping devices are not effective enough in these cases, when the column is operated at too high a temperature. It can be concluded from Fig. 6 and 7 at which oven temperatures the perfect trapping of even very volatile species, such as C_5-paraffins, which require a temperature of -80 $^\circ$C for trapping, can be achieved. Merging of the single peaks of two subsequently injected C_5 - C_{11} paraffin samples after trapping and

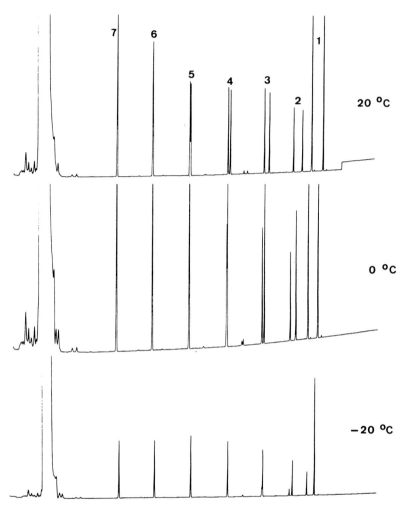

PEAK FOCUSSING (AT SPLIT SAMPLING) OF VOLATILE SAMPLE COMPONENTS BY
COOLING OF THE ENTIRE COLUMN

Two consecutive injections (60 sec) of a C_5-C_{11} hydrocarbon solution (solvent methylnaphthalene).

Effectivity of trapping in dependency on the column temperature, concluded from the overlapping
of the peak pairs of two consecutive injections.

Sample   : 0.4 µl C_5-C_{11}, diluted in methylnaphthalene
                1 pentane, 2 hexane, 3 heptane, 4 octane, 5 nonane, 6 decane, 7 undecane

Column   : 50 m methylpolysiloxane OV 1

Temperature : split mode introduction of sample at temperature specified in chromatogram.
                ballistic increase of temp. up to 20 °C, temp. progr. 20 - 200 °C, 6 °C/min

Carrier gas   : hydrogen, 13 cm/sec   Instrument   : Siemens Sichromat 2

G. Schomburg, Max-Planck-Institut für Kohleforschung, Febr. 1983

Fig. 6

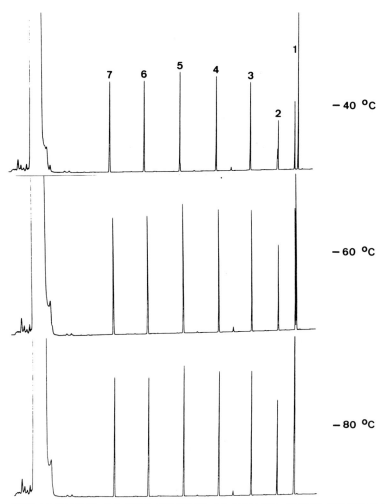

PEAK FOCUSSING (AT SPLIT SAMPLING) OF VOLATILE SAMPLE COMPONENTS BY
COOLING OF THE ENTIRE COLUMN

Two consecutive injections (60 sec) of a C_5-C_{11} hydrocarbon solution (solvent methylnaphthalene).

Effectivity of trapping in dependency on the column temperature, concluded from the overlapping
of the peak pairs of two consecutive injections.

Sample　　　:　0.4 μl C_5-C_{11}, diluted in methylnaphthalene
　　　　　　　　　1 pentane, 2 hexane, 3 heptane, 4 octane, 5 nonane, 6 decane, 7 undecane

Column　　　:　50 m methylpolysiloxane OV 1

Temperature　:　split mode introduction of sample at temperature specified in chromatogram.
　　　　　　　　　ballistic increase of temp. up to 20 °C, temp. progr. 20 - 200 °C, 6 °C/min

Carrier gas　:　hydrogen, 13 cm/sec　　　　　　Instrument　:　Siemens Sichromat 2

G. Schomburg, Max-Planck-Institut für Kohleforschung, Febr. 1983

Fig. 7

re-volatalization within the second column is only achieved by perfect trapping within the second column. High volatile compounds require very low trapping temperatures. In practice difficulties may arise when the second column is to be operated isothermally, i.e., at a temperature that is suitable for the elution of all the solutes of a certain selected cut within a reasonable separation time. The initiation of the elution from the second column is done simply after having started the temperature program. By quick (ballistic) programs, the temperature necessary for the isothermal elution or the starting temperature of a slower temperature program, scheduled for the separation, can be reached fast and with sufficient reproducibility. This fact can be concluded from the data in Fig. 8.The absolute standard deviations of the determined Kovats indicies are ranging from 0.05 - 0.1 index units. The trapping of the transferred material must be complete and reliable because for the determination of the retention indices the necessary paraffin standards must be trapped together with and in exactly the same spot of the second column after a previous injection of these paraffins into the first column. This is necessary for achieving the same starting points for the elution of sample components and standards. In one of our previous set-ups which was already described in 1975, we had arranged a second injector between the first and the second column in order to be able to introduce the necessary standard compounds[26], i.e., without a pre-separation, as practiced in this particularly work. For the sake of identification, isothermally determined retention indices are to be preferred for the reasons discussed above, but the trapping of the transferred cuts may become difficult like in the case of more volatile solutes (< C_{10} paraffins) if higher column temperatures are to be maintained. "Linear" retention indices can only be used for interlaboratory comparisons, when they were measured under defined conditions, i.e., by standardization of the enumerated significant system parameters. Simple comparisons of linear with isothermal retention indices, as shown in Fig. 9, can only be done when the temperature dependency of the isothermal retention index is known, which happens to be not the case, if the retention data of unknowns are to be determined. Correction of linear retention indices for temperature dependency can approximately be

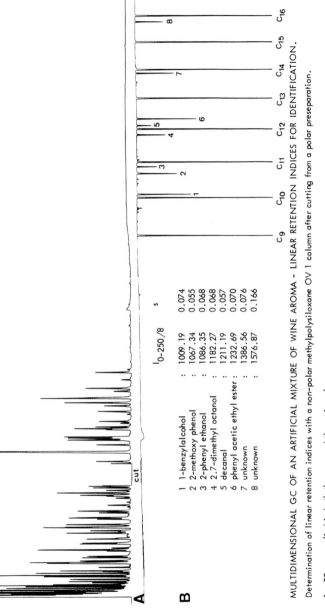

137

$I_{0-250}/8$	s
1 1-benzylalcohol :	1009.19	0.074
2 2-methoxy phenol :	1067.34	0.055
3 2-phenyl ethanol :	1086.35	0.068
4 2.7-dimethyl octanol :	1182.27	0.068
5 decanol :	1211.19	0.057
6 phenyl acetic ethyl ester :	1232.69	0.070
7 unknown :	1386.56	0.076
8 unknown :	1576.87	0.166

MULTIDIMENSIONAL GC OF AN ARTIFICIAL MIXTURE OF WINE AROMA - LINEAR RETENTION INDICES FOR IDENTIFICATION.

Determination of linear retention indices with a non-polar methylpolysiloxane OV 1 column after cutting from a polar preseparation.

1. TP was applied in both the pre- and the main column.
2. Trapping by cooling of entire column and sectional cooling of column inlet.
3. Standard paraffins were trapped from a preceding separation together with the cut components.
4. Standard deviations of repeatability were determined.

Sample   : Mixture of about 120 components occuring in wine aroma

Column   : **A** 25 m Ucon 75 H 90000, 0.27 mm i. d.   **B** 50 m methylpolysiloxane OV 1 , 0.27 mm i. d., d_f = 1.5 µm

Temperature : **A** 80 °C, 6 °C/min to 240 °C   **B** 0 °C, 5 min iso, to 250 °C, 8 °C/min

Carrier gas : H$_2$ Instrument : Double oven, Siemens Sichromat 2   Analysis time : 40 min

G. Schomburg, H. Husmann, L. Podmaniczky, F. Weeke, Max-Planck-Institut für Kohlenforschung, August 1983

Fig. 8

achieved when assumptions about the chemical structure, or the polar functional groups are possible. The temperature dependencies $\partial I/\partial T$ of the retention indices of flavor compounds are, however, strongly dependent on their chemical structure as can be seen from the data of the Table in Fig. 9. Consequently the influence of the programming parameters, the flow rate, and the stationary phase content on the "linear" retention indices is very strong if the $\partial I/\partial T$'s of the Kovats indices are very high. In the Table of Fig. 10 the comparison of the various types of standardized retention data as well as the retention temperatures measured within the same column illustrate the problem of retention temperature dependency regarding the practical use of retention indices. For compounds of low $\partial I/\partial T$'s: alcohols, esters, etc., fewer problems exist for the comparison of the data. According to the difficulties described regarding the determination of isothermal retention data and because of the problems arising with linear index usage for Table and literature search identification work, the author expresses doubts whether (especially in the inter-laboratory mode) chromatographic retention data as obtained from multidimensional set-ups, operated with temperature programming can be used successfully in practice. The former difficulties arising with irreproducible polarity of the columns used do no longer exist, when the data serving for identification and characterization purposes are being determined with the new types of low polarity alkylpolysiloxane columns which can be reliably produced nowadays. Overlappings of important components by broad peaks originating from column overloading, especially in the case of highly polar solutes can easily be abolished by multidimensional GC transferring the adequately selected cuts from a polar pre-column to the non-polar column, which only contain a limited number of components. Especially the gas chromatographic discovery of trace components can be extremely facilitated even when such traces are going to be overlapped by major components. The improvement of resolution is highest when the overlapping species have a different polarity. Characterization of trace components in such a manner resolved from major components can then also be done by retention data: The non-polar methylpolysiloxane columns should be preferred since their polarity is very stable and reproducible.

"ISOTHERMAL" AND "LINEAR" RETENTION INDICES OF DIFFERENT CHEMICAL COMPOUNDS.

Dependency of temperature or type of temperature program.

Stationary phase : methylpolysiloxane OV 1, crosslinked (H 1713 a)*

		"isothermal" (Kovats)		"linear"	
		I_{100}	$\partial I / \partial T$	$I_{70 \rightarrow 250/4}$	$I_{90 \rightarrow 200}\ °C/4\ °C/min$
1-Hexanol	:	849.73	- 0.62	845.52	843.36
Decalin	:	1058.69	+ 5.76	1049.12	1055.54
2-Nonanol	:	1085.37	- 0.29	1085.05	1083.35
Naphthalene	:	1161.16	+ 6.53	1157.28	1163.25
Heptylpropionate	:	1187.26	- 0.71	1186.96	1185.70

*manufacturing code

G. Schomburg, H. Husmann, L. Podmaniczky, F. Weeke, Max-Planck-Institut für Kohlenforschung, August 1983

Fig. 9

ISOTHERMAL AND LINEAR RETENTION INDICES AS WELL AS RETENTION TEMPERARURES OF DECALIN.

Obtained with different types of temperature programs; isothermal indices at different temperatures.

Stationary phase : methylpolysiloxane OV 1, crosslinked (H 1713 a)*

Temp. program :	70-200/1	90-200/1	70-200/2	90-200/2	70-200/4	90-200/4	90-200 °C/8 °C/min
± Linear index :	1039.52	1048.29	1043.68	1051.31	1049.12	1055.54	1061.77
Retention temperatures :	81.1	96.9	89.6	102.8	102.4	113.2	130.8 °C
	80 °C	100 °C	120 °C	130 °C		$\partial I / \partial T$	
Isothermal index :	1047.64	1058.69	1070.15	1076.55		+ 5.75	

*manufacturing code

G. Schomburg, H. Husmann, L. Podmaniczky, F. Weeke, Max-Planck-Institut für Kohlenforschung, August 1983
Fig. 10

The most important method for identification of the components of complex mixtures is, of course, the GC-MS technique[25] which combines the high selectivity of separation with the very high specifity of detection (response). The major limitations of combined MS usage in high resolution GC of flavors as well as aromas are:

- insufficient information content of even the entire mass spectrum for the identification of unknowns and
- non-availability of spectra within the libraries
- lack of detection sensitivity for components of very low concentrations
- inadequate computer software for the library search
- unreliable spectra as contained within the libraries available.

The efficiency and effectiveness of modern library search programs is demonstrated by the results obtained from GC-MS of peppermint oil (origin Formosa). A similar peppermint oil had been analyzed by combined GC-FTIR and GC-MS by Wilkins et al.[31]. We were able to identify the major as well as most of the minor components of such mixture by using the software developed by D. Henneberg et al.[32] which seems to be superior to the Bieman software and the library used by Wilkins. More components and such of lower concentration could be identified reliably because of the wide dynamic range of mass spectrometric signal processing. With our GC-MS-analysis we were able to obtain good mass spectra from low concentration components (0.1 %) which could be used for library search identification. (See Fig. 11). The spectrum with the sequence number 162 was obtained from a component which was contained with the sample at a concentration of < 0.1 %. The spectrum was good enough to identify the compound trans-linaloloxide by library search. There was no urgent need for IR information in order to attain a reliable identification, in this case selected as an example of GC-FTIR application. This fact can, for example, also be concluded from the comparison of mass spectra of α-and β-pinene which Wilkins considers to be not characteristic enough for identification. Of course, it is not always possible to distinguish between certain isomeric or diastereomeric species which are easier to characterize by using the IR spectra. At present, lack of sensitivity for low concentration components the additionally high expenditures for FTIR

142

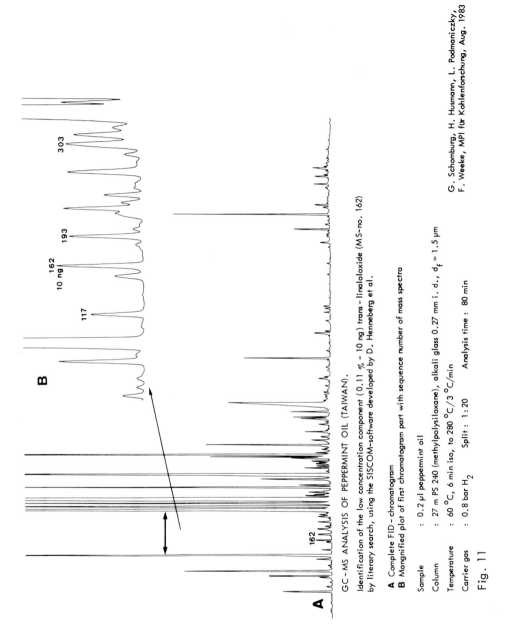

GC - MS ANALYSIS OF PEPPERMINT OIL (TAIWAN).

Identification of the low concentration component (0.11 % – 10 ng) trans-linaloloxide (MS-no. 162) by literary search, using the SISCOM-software developed by D. Henneberg et al.

A Complete FID – chromatogram
B Magnified plot of first chromatogram part with sequence number of mass spectra

Sample	: 0.2 μl peppermint oil
Column	: 27 m PS 240 (methylpolysiloxane), alkali glass 0.27 mm i. d., d_f = 1.5 μm
Temperature	: 60 °C, 6 min iso, to 280 °C/3 °C/min
Carrier gas	: 0.8 bar H_2 Split : 1:20 Analysis time : 80 min

Fig. 11

G. Schomburg, H. Husmann, L. Podmaniczky,
F. Weeke, MPI für Kohlenforschung, Aug. 1983

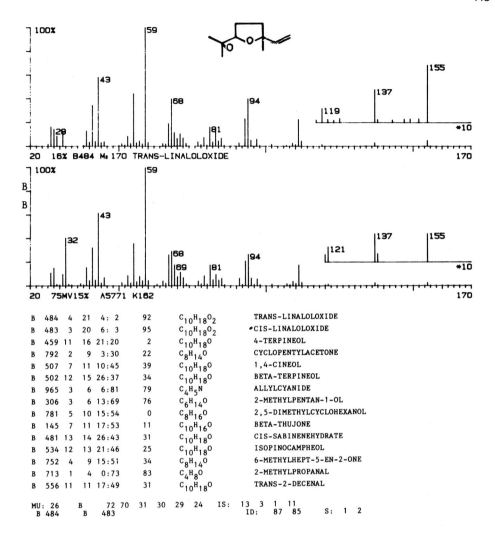

B	484	4	21	4: 2	92	$C_{10}H_{18}O_2$	TRANS-LINALOLOXIDE
B	483	3	20	6: 3	95	$C_{10}H_{18}O_2$	·CIS-LINALOLOXIDE
B	459	11	16	21:20	2	$C_{10}H_{18}O$	4-TERPINEOL
B	792	2	9	3:30	22	$C_8H_{14}O$	CYCLOPENTYLACETONE
B	507	7	11	10:45	39	$C_{10}H_{18}O$	1,4-CINEOL
B	502	12	15	26:37	34	$C_{10}H_{18}O$	BETA-TERPINEOL
B	965	3	6	6:81	79	C_4H_5N	ALLYLCYANIDE
B	306	3	6	13:69	76	$C_6H_{14}O$	2-METHYLPENTAN-1-OL
B	781	5	10	15:54	0	$C_8H_{16}O$	2,5-DIMETHYLCYCLOHEXANOL
B	145	7	11	17:53	11	$C_{10}H_{16}O$	BETA-THUJONE
B	481	13	14	26:43	31	$C_{10}H_{18}O$	CIS-SABINENEHYDRATE
B	534	12	13	21:46	25	$C_{10}H_{18}O$	ISOPINOCAMPHEOL
B	752	4	9	15:51	34	$C_8H_{14}O$	6-METHYLHEPT-5-EN-2-ONE
B	713	1	4	0:73	83	C_4H_8O	2-METHYLPROPANAL
B	556	11	11	17:49	31	$C_{10}H_{18}O$	TRANS-2-DECENAL

```
MU: 26     B      72 70  31  30  29  24   IS: 13  3  1  11
   B 484   B     483                       ID:   87 85      S:  1  2
```

LIBRARY SEARCH OF MASS SPECTRUM

Mass spectrum of peak no. 162 in GC-MS analysis of peppermint oil compared to library spectrum of trans-linaloloxide.

Hit list of reference spectra ranked for "similarity".

Fig. 12

equipment and the complicated infrared cell geometry with its large volume and the corresponding interfacing problems may prevent an extended use in practice, especially in combination with GC-MS set-ups which are already in use in the common laboratories.

Retention data can be successfully used as well in addition to the mass spectrometric information in order to support the final decision about the identity of an unknown and repetitive accumulation of pre-separated components may help to solve these problems in the case of low concentration components. Combining a multidimensional high resolution system with a modern mass spectrometer and an adequate computer software for library searching as well as reliable and extensive libraries for mass spectra can, of course, help to improve the identification of even low concentration components in very complex mixtures. This could simply be demonstrated in our work through transfer of the components contained in the special cut after trapping within the second column of the MDGC system to the mass spectrometer. This was achieved by disconnecting the second (alkylpolysiloxane) column and reconnecting it to the interface of a mass spectrometer. Generally, the elucidation of the chemical structure and the configuration of unknown constituents would be much easier, if NMR spectra of the various nuclei including ^{13}C could be obtained from such species that can only be separated from other species by high resolution. The sample capacity of high resolution (capillary) columns even of such with very thick films ($\sim$ 5 μ) is so low, however, that not enough material can be obtained from such separations, except with an accumulation from very many repetitive separations, as has been described for ^{1}H-NMR application by Roeraade[33]. For the ^{13}C-NMR this approach does not seem to be very promising because of the lack of sensitivity even when accumulating the intensity from a very large number of pulses transients using the Fourier technique. There is no doubt that the sample capacity of columns under conditions that give rise to the efficiency just necessary for the separation of certain (mainly) isomeric or diastereomeric species can be increased to some extent by coatings with thicker films or by a certain increase of the internal diameter; of course, the most effective approach for the achievement of higher resolution is outmost selectivity optimization via temperature and/or column polarity since higher column loads of the significant components are possible. Despite from these measures of optimization repetitive (cyclic) separations with accumulation of the interesting fractions in special cold traps seems to be the only way for getting sufficient material for the application of even Fourier NMR. It should also be reconsidered if not packed micro particle columns could solve such problems better than capillary columns although special support materials have to probably be developed and high pressure drops as well as the technical problems related have to be faced thereby. Concerning the use of infrared spectroscopy, problems of sensitivity arise for low concentration components with only a few characteristic intensive IR absorptions. In this case problems also arise with the dead-volume of the infrared cells and the necessary connection lines between column outlet and light pipe. Moreover, adsorption problems arise on the surfaces of the light pipe at too low operation temperatures. Nevertheless, it seems practical to make use of FTIR-identifications or characterizations, especially if an adequate approach of scale-up of the sample capacity of the GC-separation via phase-ratio, film thickness i.d. of column, and optimization of selectivity is adopted. Excessively strong decrease of efficiency by such provisions has to, of course, be avoided.

If many of the Fourier pulses must be applied for achieving sufficient sensitivity, the stop-flow-technique must be applied in order to keep the interesting section of the eluate longer within the cuvette. The dead volumes of the light pipe and the connection lines have to be properly compensated by make-up-gas introduction. A study on the relations between volume of lightpipes and GC-FTIR-sensitivity was published by Wilkins[34]. Because the IR cell operates like a concentration dependent detector, make-up gas-flows may not be too high because the sensitivity of IR measurements in the eluate decreases proportionally to the dilution within the carrier gas. At present, there is no doubt that major components of an aroma suitable IR spectra may be achievable, but the necessary amount of substance seems to be in the range of ~ 50 ng even for compounds of intensive IR absorption. Limitations in the application of IR for "hyphenated GC"-identification are due to:

146

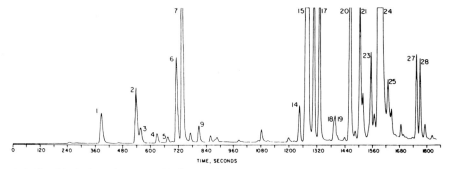

Flame ionization detector trace for peppermint oil separation.

GC/FT-IR/MS Analysis of Peppermint Oil

peak	component	GC peak area[a]	IR search	MS search
1	α-pinene	14	+	+
2	β-pinene	2.3	+	−
3	sabinene	0.5	N[b]	−
4	myrcene	0.3	N	+
6	l-limonene	3.0	+	+
7	eucalyptol	7.2	+	+
14	trans-sabinene hydrate	1.0	--	−
15	l-menthone	20.1	+	+
17	d-isomenthone	5.7	+	+
20	menthyl acetate	7.0	+	−
21	neomenthol	3.6	+	+
22	terpinene-4-ol	1.2	−	−
23	β-caryophyllene	2.8	−	−
24	l-menthol	29.0	+	+
25	pulegone	3.5	−	−
26	α-terpineol	0.8	−	−
27	germacrene	2.0	−	−
28	piperitone	1.3	−	+

[a] Uncorrected for differing FID response. [b] No infrared spectrum obtained, due to inadequate sensitivity.

GC/FT - IR/MS ANALYSIS OF PEPPERMINT OIL

Charles L. Wilkins et al., Anal. Chem. 54 , 2260-2264 (1982)
Fig. 13

- a too low amount of component concentration in the maximum admissible sample volume
- a too low sample capacity of the column used at the desired efficiency
- too low intensity of IR absorption of the compounds of interest
- a too poor instrument sensitivity at the maximum number of pulses, acceptable for the speed of the chromatogram
- a too high scavenge gas-flow for dead-volume compensation
- an uncharacteristic IR spectrum.

CONCLUSION

Multidimensional GC (in coupled columns) and hyphenated techniques such as GC-MS or GC-IR help either to improve resolution, to shorten analysis time, to decrease detection limits, or to identify known or unknown species of complex mixtures. In case low concentration components which may be of olfactoric significance, for example, the sensitivity of the hyphenated infrared and in some cases also the mass-spectrometers may not be high enough for identification or structure elucidation. Considerations about sample capacity at the minimally required separation efficiency and resolution are important. Non-polar thick film columns are proposed to be used in particularly with multidimensional systems when using a polar column for the preseparation. Independent temperature or temperature programed optimization of each single column is necessary when using coupled columns for this type of work. Highly reproducible retention data of cut-components that are no longer overlapped, can be measured reproducibly in MDGC set-ups using a multi-oven instrument. The problems arising with the different temperature dependencies of retentions on the linear retention indices can only be eliminated by applying standardized temperature programs regarding initial temperature, program rate and carrier gas flow to columns of defined but preferably thick film capillary columns.

148

References

1. Schomburg, G., Behlau, H., Dielmann, R., Weeke, F., Husmann: J. Chromatogr. 142, 87 (1977)

2. Grob, K., Grob, K. Jr.: Chromatogr. -151, 311 (1978)

3. Galli, M., Trestianu, S., Grob, K. Jr.: HRC & CC 2, 366 (1979)

4. Schomburg, G.: Proceedings IV. Int. Symposium on Capillary Chromatography Hindelang 1981, Hüthig Verlag Heidelberg, p. 371 and A 921

5. Poy, F., Visani, S., Terrosi, F.: J. Chromatogr. 217, 81 (1981)

6. Schomburg, G.: Proceedings V. Int. Symposium on Capillary Chromatography Riva del Garda 1983, Elsevier Amsterdam, p. 280

7. Grob, K. Jr.: Proceedings V. Int. Symposium on Capillary Chromatography Riva del Garda 1983, Elsevier Amsterdam, p. 254

8. Grob, K., Grob, K. Jr.: J. Chromatogr. Sci. 7, 584 (1969)

9. Grob, K., Grob, K. Jr.: J. Chromatogr. 94, 53 (1974)

10. Grob, K., Grob, K. Jr.: HRC & CC 1, 57 (1978)

11. Grob, K., Grob, G., Grob, K. Jr.: J. Chromatogr. 211, 243 (1981)

12. Sandra, P., Redant, G., Schacht, E., Verzele, M.: HRC & CC 4, 411 (1981)

13. Wright, B.W., Peaden, P.A., Lee, M.L., Stark, T.: J. Chromatogr. 248 17 (1982)

14. Blomberg, L., Buijten, J., Markides, K., Wännman, T.: HRC &CC 4, 578 (1981)

15. Schomburg, G., Husmann, H., Ruthe, S., Herraiz, M.: Chromatogr. 15, 599 (1982)

16. Schomburg, G.: J. Chromatogr. Sci. 21, 97 (1983)

17. Chrompack, Middelburg, Various Advertisements

18. Grob, K., Grob, G.: HRC & CC 2, 31 (1979)

19. Grob, K., Grob, G.: HRC & CC 3, 197 (1980)

20. Schomburg, G., Husmann, H., Behlau, H.: Chromatographia 13, 321 (1980)

21. Schomburg, G., Husmann, H., Behlau, H.: J. Chromatogr. 203, 179 (1980)

22. Blomberg, L., Markides, K., Wännman, T.: Proceedings V. Int. Symposium on Capillary Chromatography Riva del Garda 1983, Elsevier Amsterdam, p. 73

23. Markides, K., Blomberg, L., Buijten, J., Wünneman, T.: Proceedings V. Int. Symposium on Capillary Chromatography Riva del Garda 1983, Elsevier Amsterdam, p. 117

24. Dandeneau, R., Zerenner, E.H.: HRC & CC 2, 351 (1979)

25. Schomburg, G., Henneberg, D.: Analysensysteme und Methoden, GIT-Verlag Darmstadt, 2. Auflage p. 119 (1982)

26. Munari, F., Trestianu, S.: Proceedings V. Int. Symposium on Capillary Chromatography, Riva del Garda 1983, Elsevier Amsterdam, p. 327

27. Schomburg, G., Weeke, F.: Chromatographia 16, 87 (1982)

28. Schomburg, G., Hübinger E., Husmann, H., Weeke, F.: Chromatographia 16, 228 (1982)

29. Schomburg, G., Husmann, H., Behlau, H., Bastian, E., Weeke, F.: Proceedings V. Int. Symposium on Capillary Chromatography, Riva del Garda 1983, Elsevier Amsterdam, p. 381, to be published in HRC & CC, 1983

30. Schomburg, G., Husmann, H., Weeke, F.: J. Chromatogr. 112, 205 (1975)

31. Wilkins, Ch. L., Giss, G. N., White, R. L., Brissey, G. M., Onyiriuka, E. C.: Anal. Chem. 54, 2260 (1982)

32. Henneberg, D.: Advances in Massspectrometry, Editor: A. Quayle, 8 B, p. 1511 (1979), Institute of Petroleum, Heyden & Sons, London (1979)

33. Roerade, J., Enzell, C. R.: HRC & CC 2, 123 (1979)

34. Wilkins, Ch. L., Giss, G. N.: Abstracts of papers presented at the Pittsburgh Conference 1983, Atlantic City, paper no. 605

CONCENTRATION AND GC-MS ANALYSIS OF TRACE VOLATILES BY SORP-
TION-DESORPTION TECHNIQUES

S. Nitz and E. Jülich
Institut für Lebensmitteltechnologie und Analytische Chemie
Technische Universität München
D-8050 Freising-Weihenstephan

Introduction

Qualitative and quantitative determination of trace organics
present in solution and gaseous media are of great importance
in pollution control, aroma research and biochemical-chemo-
reception studies. Scientific findings about the composition
of natural products by example have shown that it is necessary
to penetrate in the domain of traces in order to be able to
characterize an aroma, since in many cases their sensoric contribution is of essential character (1).

Although gas chromatography has permitted major advances in
the field of volatile analysis, the study of the volatile composition of diluted vapour systems (eg. headspace, air) or complex samples from natural products containing nonvolatile
materials confronted the investigators with additional problems.

As to the analysis of diluted vapour samples a direct
injection can only be used for the detection of major components with high vapour pressure; injection of concentrated
liquid samples is problematic since capillary columns with
high efficiency start to overload in the range of approximately
100 ng per single component. These limits are reached quickly,
and if the sample also contains larger amounts of nonvolatile
materials, a rapid deterioration of column performance might

occur. This is a typical problem that arises in the analysis
of complex mixtures that contain components with wide range of
concentrations, polarities and volatilities.

Thus, for the analysis of trace organics in vapour samples an
adequate concentration step preceeding the analytical determination is necessary. Concentration and fractionation is
required if trace organics in liquid extracts have to be analyzed.

These concentration and analysis systems must be specially
designed to permit short sampling and analysis cycle times.

Concentration and Analysis of Gaseous Samples
by Glass-Capillary Gas Chromatography

The analysis of gaseous samples by gas chromatography has been
widely applied in air pollution control as well as in food and
flavour chemistry.

Several methods have been used for concentration. Cryogenic
techniques (eg. cold traps, freezing of sample in empty capillaries, direct introduction in cooled open tubular columns),
adsorption on solid adsorbents (charcoal, silicagel, alumina,
porous polymers) and conventionally coated GC supports have
been applied (2-7).

On column cryogenic concentration is technically difficult
with large volumes of gaseous samples and is limited to stationary phases which maintain their liquid state at low temperatures (8).

Problems with water interference and solvent impurities have
been experienced with freezing of sample in empty tubes and
solvent elution (9).

The best results have been obtained by displacing large volumes of gaseous samples through appropriate adsorbing devices, retaining most organic volatiles. Recovery of the trapped
volatiles can be achieved by elution with solvents or by
thermal desorption. The approach using a concentration preco-

lumn and subsequent thermal desorption seems to be most promising.

In terms of relative inertness of the adsorbent, the application of porous polymers and support bonded chromatographic phases as trapping material have proved to be the most suitable for thermal desorption. Polymers were used at room temperature, and water did not represent a particular problem since they are nonpolar and have little affinity for water. Tenax GC, a 2,6-diphenyl-p-phenylene-oxide polymer as a medium for trapping and transfer of volatile organics to a gas chromatograph has shown to be the most appropriate.

A general problem is the transfer of thermal desorbed compounds onto a capillary column. Commercial equipment for transferring on packed columns is available, but their design is incompatible with the requirements of capillary columns. As high resolution is demanded, a narrow band of desorbed material has to be deposited immediately before the column and vapourization has to occur in a very short time at the start of the chromatographic run.

Many sophisticated approaches have been published in literature. The main drawback of them is that for their realization a series of greater modifications and additional equipments are necessary, besides the fact that in many cases switching valves and metallic transfer lines were located in the pathway of the components (10, 11). For this reason a system for entraining thermal desorbed substances onto a capillary column without the need of splitting and greater modifications of the chromatographic system was developed. This system will be described in the following.

1. Desorption oven and adsorbing cartridges

The principle of operation of the desorption manifold is
that of negative temperature gradient described by Kaiser
(12). The desorption oven was mounted directly on top of
the injection port, and on the upper end it was provided
with a platinum sensor for air temperature regulation. Air
was heated by means of a hot pistol; the heating cartridge
was connected to the temperature regulation device. The
Tenax cartridges can be heated from room temperature up to
250 ^{O}C in about 2 - 3 minutes.

The adsorbing cartridges used were made of glass tube
(1/4" O.D., 4 mm ID, 1o cm length), packed with 150 mg
Tenax GC 80-100 mesh (Alltech Europe Inc.) and plugged at
both ends with silanized glass wool. Preconditioning was
done at 250 ^{O}C with a flow rate of 30 ml/min of purified
nitrogen during 25 hrs. The upper end was fitted with a
gas connector, the lower end with an injection needle.

2. Trapping of desorbed volatiles

For intermediate trapping of the desorbed volatiles a
slightly modified programmable temperature vaporizer (PTV)
was used (Fig. 1). This injector is normally cooled with
air and can be heated up to 250 ^{O}C in about 15 - 20 seconds.
This operation is automatically started by the microprocessor of the GC (Dani, Mod 6800). A glass capillary tube
(85 mm long, 1.6 mm O.D., 0.6 mm I.D.) is located inside
the body of the injector which is filled with silanized
glass wool. The precolumn is easily removable without disconnecting the main capillary column. For thermal desorption
the following modifications were carried out. The capillary

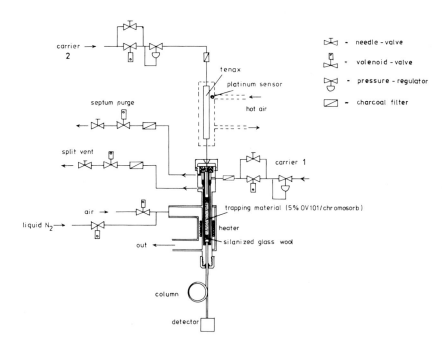

Figure 1   Programmable temperature vapourizer for intermediate trapping.

glass insert was filled with 5 % OV-101 on Chromosorb W and
plugged at both ends with silanized glass wool. Additionally,
a solenoid valve was fitted to the cooling pipe in order to
be able to cool with liquid nitrogen. Thus, during the preconditioning phase the injector cools down to 50 $^{\circ}$C with
air and switches over automatically to liquid N_2 cooling
until - 150 $^{\circ}$C are reached. A control module provides the
means for setting the initial and final PTV temperatures,
and for opening and closing the control valves of the splitter. This modification does not disturb the normal use of
the gas chromatograph, and an additional advantage for normal
operation is that non-volatile residues are retained by the
trapping material and do not enter the capillary column.

3. System operation

The whole analysis is carried out under the control of the
gas chromatograph microprocessor, and involves the following
steps (see Fig. 1 for details).

1. Installation of Tenax trap in desorption oven;
2. Trap cooling, wait for temperature to be -150 $^\circ$C;
   Set GC-oven at initial temperature;
3. Insert needle of Tenax trap in injector;
4. Carrier gas (2) on, carrier gas (1) off, splitter on;
5. Start desorption-oven heating;
6. Wait for sample transfer into cold trap;
7. Carrier gas (2) off, carrier gas (1) on, split off;
8. Heat injector from -150 $^\circ$C to 220 $^\circ$C, start GC-program;
9. End of run, go to stand by.

4. System performance

Table 1   Standard deviation of peak area of model
          compounds for thermal desorption from
          Tenax GC

Substance		Area	Standard Deviation	%
Pentane	(a)	15488	2382	15.4
Hexane	(a)	17354	3381	19.5
Heptane	(a)	17516	353	2.0
Octane	(a)	9145	435	4.8
Nonane	(a)	7981	343	4.3
Limonene	(b)	9566	723	7.6
Linalool	(b)	11476	784	6.8
Caryophyllene	(b)	11677	1152	9.9
Humulene	(b)	13620	2041	15.0

(a) gaseous   (b) in solution

The degree of reproducibility of the analytical system was
determined by measuring the area of peaks in 6 replicated
runs (Table 1). The standard deviation for thermal desorption is within a reasonable order of magnitude, keeping in
mind, that for the thermal desorption these values represent the total variability of the procedure. The relationship between sample size and peak area of a model sample in
the range of 10 to 100 ng for 4 substances is shown in
Fig. 2. A good linear correlation can be observed, even at
a concentration of 100 ng per component in the mixture.

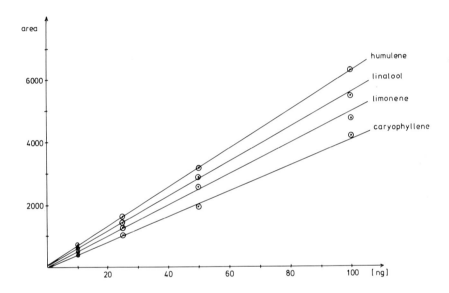

Figure 2   Graphs of chromatographic peak area against
            sample size for selected compounds sampled
            on Tenax-GC.

Experience with headspeace analysis of beer wort has shown
that the trapping capacity of this intermediate trap is in
the range of several micrograms. A typical chromatogram
obtained from a headspace of beer wort is shown in Fig. 3.

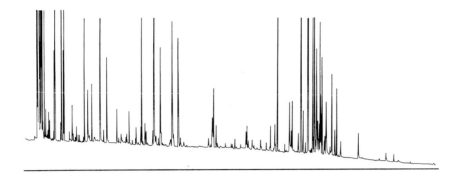

Figure 3   Chromatogram of 50 ml beer wort headspace
            desorbed from Tenax-GC; 0.25 mm x 30 m quartz
            capillary coated with SE-30; program from
            0° - 200 ^{o}C at 2 ^{o}C/min.

In this case 50 ml of headspace were adsorbed on Tenax-GC
and after desorption and intermediate trapping capillary
separation was performed. It was verified that the introduction of the trapping material into the capillary insert
did not introduce any deterioration of the retention times
and the column performance, in comparison to chromatograms
obtained with normal GC equipment.

The intrinsic problems associated with enrichment of gaseous
samples by adsorption techniques are not eliminated with
this system. Proper choice of adsorbing material, artifact
formation due to high desorption temperature, decomposition
of unstable compounds, low desorption recoveries, represen-
tativity of the analysed vapour phase have still to be considered. A wide range of compounds can be concentrated and
analyzed by this method; the interpretation of data as for
absence of particular compounds or quantity estimation must
be made with caution.

At this point, a new system, based on desorption by means
of microwave application, remains to be mentioned (13).
High speed desorption is possible with microwaves, but their

application up to now is limited to the use of activated
charcoal. Formation of artifacts on this type of adsorbing
material has been described in literature and is also
observed when microwave desorption is applied.

Trace analysis of complex mixtures

Although enrichment on adsorbing devices is a valuable tool for
the detection of substances present in gaseous media, it is not
appropriate for the determination of traces in a complex matrix
such as aroma extracts of natural products. Since the contribution of trace compounds to total vapour pressure is negligible,
they can only be detected by analysis of the liquid extract.
Large sample sizes are required if traces are to be determined
and column overloading by major components is reached quickly
with subsequent deterioration of column performance. The con-
ceptionally simplest method of resolving a compound from a
complex sample is: chromatography in a first column, trapping
of the fraction of interest and re-injection into a second
column. This off-line method is less suited for small quantities and prone of artifacts, besides the fact that it is time
consuming if several components have to be analyzed.

If it is possible however, to introduce on-line into a second
analytical column only the substances to be analyzed and vent
the part which causes overloading, the problem can be overcome
to a large extent. This can be advantageously performed by
using double column systems with packed precolumns. A number
of versatile two-dimensional gas chromatographic systems were
described by Schomburg et al.(14-17), who have done pioneering
work in this field. Two dimensional chromatography serves three
primary purposes: (a) to effectively resolve complex mixtures,
(b) to make use of retention data for identification purposes
and (c) sampling and enrichment of trace amounts in dilute
solutions.

The sampling and enrichment of trace amounts is performed with packed pre-columns because of their high sample capacity. The substances of interest are intermediately trapped after pre-separation on the packed column and after trap heating; the separation of the interesting species is performed in the main capillary column. If necessary repetitive pre-separation for enrichment of trace components can be carried out. Intermediate trapping is achieved by cooling the first centimeters of the main capillary column. Up to now this technique was reserved to research laboratories with many years of experience in this field. Instruments had to be specially designed for multidimensional gas chromatography and required a multitude of functional elements such as valveless switches, intermediate traps, dual ovens, automatic timers and others. These may be the reason why its application was mainly restricted to specialized research laboratories. Fortunately a chromatographic device which incorporates all the hardware and software for multi-dimensional gas chromatography is nowadays commercially available. By means of this system a total transfer from packed to capillary columns is possible. Intermediate trapping in this case is achieved by sorption of the compounds of interest on polymers or GC support materials. In its standard configuration the trap is filled with 5 % SE 30 on Chromosorb G 60-80 on a length of approximately 10 mm. That such a combination is a valuable tool for aroma research is out of question. About the potentialities of this analytical device for aroma research will be reported.

System description and operation

Figure (4) shows a schematic of the total transfer system (Siemens Mod. Sichromat 2). The gas chromatograph has two ovens thermally insulated which are independently temperature programmable. The intermediate trap is fitted to the outlet of the packed column and to the inlet of the capillary. The

161

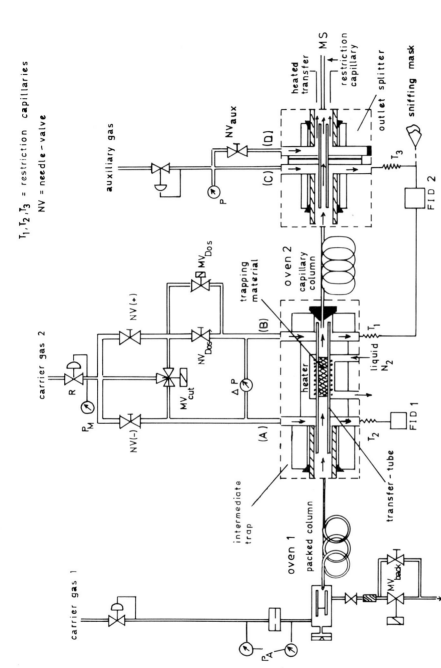

Figure 4   System for total transfer from packed to capillary columns with intermediate trapping and device for parallel MS/sniffing or MS/FID registration (Siemens, Mod. Sichromat 2).

direction of substance flow emerging from the packed column
depends on the pressure difference Δ P between points (A) and
(B). The magnitude of differential pressure is adjusted by
means of needle-valves NV(-), NV(+),and NV dos.

The chromatographic run can be divided in three steps:

1. Cut: all substances eluting from the packed column are
   transferred quantitatively to FID-detector (1)
   (ΔP between (A) and (B) is negative; the adequate
   value has to be determined experimentally). If the
   adjustment of pressure difference is not correct,
   substance will flow through transfer tube and T_1 to
   FID-detector (2) and an additional peak will be
   registrated.

2. Transfer to trap: Substances eluting from the packed
   column are transferred quantitatively into the transfer tube and retained in the cooled trap (ΔP between
   (A) and (B) is positive; the adequate value has to be
   determined experimentally). If the adjustment of pressure difference is not correct, a substance peak will
   also be registrated on FID-detector (1).

3. Transfer to capillary: The transfer tube is heated up and
   desorbed substances are transferred to the capillary
   column. (The differential pressure between (A) and (B)
   is positive but lower in magnitude as the value in
   step 2. If transfer is not quantitative, two substance
   peaks will be registrated by FID-detector (2), because
   substance is splitted at the beginning of the capillary
   column.

After transferring the substances onto the capillary column the
temperature program in oven (2) is started. The substances
eluting from the capillary can be directly monitored with

FID-detector (2), or the outlet can be splitted for simultaneous MS-registration or, if a sniffing device is available, parallel sniffing/MS-detection can be performed. The outlet splitter shown in Fig. 4 functions in a similar way as the intermediate trap. The split ratio between FID (2) and mass spectrometer is dependent on the difference in pressure between points (C) and (D). This can be adjusted with needle valve (NV-aux).

Applications

1. Enrichment of selected compounds. Figure 5 shows a typical chromatogram of a selected cut of a cauliflower extract which was resolved on a capillary column.

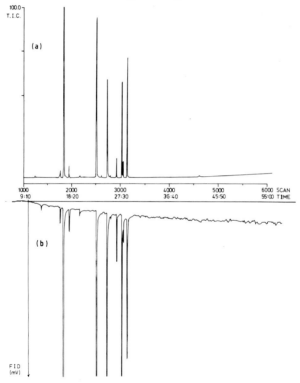

Figure 5   Comparison of gaschromatographic separation and retention times obtained with total ion current detector Finnigan 4020 (upper trace) and FID (lower trace).

164

Chromatogram (a) is the reconstructed total ion current
(TIC) and chromatogram (b) that of the corresponding
FID-(2) signal. As can be seen, peak shape is still good
and registration of TIC- and FID-signal are not time delayed.
The tailing observed on the FID-chromatogram for the major
components in this case has to be attributed to the nature
of the substances, as a capillary run (Fig. 6) with the
same column on a normal gas chromatograph showed the same
behaviour. Obviously for this sample the polarity of the
column is not the most appropriate.

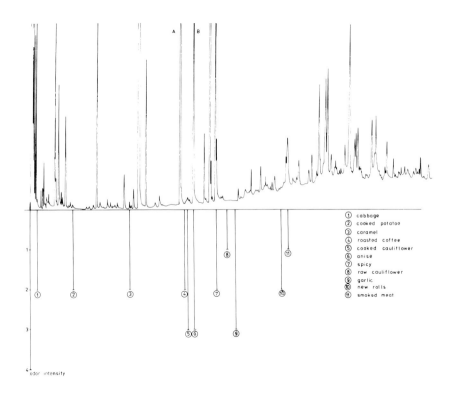

Figure 6   Chromatogram of parallel sniffing - FID
           registration (only intense odours of minor
           components are given), 0.25 mm x 30 m glass
           capillary coated with SE 54 and programmed
           from 65 - 250 °C at 3 °C/min.

In Fig. 6 we additionally registrated on an arbitrary scale
the odour intensity of substances perceived by means of
sniffing chromatography of the same cauliflower extract.
Only compounds with high odour intensity were considered.
In the region between peaks (A) and (B) an odour which was
described as "typical cooked cauliflower" was perceived
by different persons. A normal GC-MS run of the sample gave
a mass spectrum of very low purity since in the same region
other minor components were eluted.

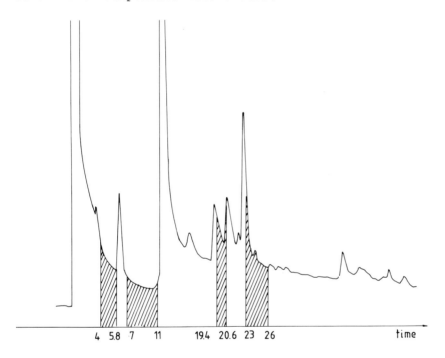

Figure 7   Chromatogram of cooked cauliflower extract
            obtained on the packed precolumn, cutting
            areas are indicated; 2 mm x 1 m packed glass
            column filled with 5 % OV-101 on Chromosorb W
            70 - 100 mesh, carrier gas helium, initial
            temperature 80 ^{O}C, programm from 70 - 280 ^{o}
            at 5 ^{O}/min.

An appropriate cut on the packed column (Fig. 7) and capillary separation cleaned up that chromatographic area (Fig. 8). Chromatogram (a) shows the TIC obtained when a cut between 19.4 to 26.0 minutes was performed. The injected amount on the capillary was approximately 1 μl. Simultaneous sniffing during MS-acquisition gave us the certainty that component (1) was the one we were looking for. The chromatogram obtained performing a cut between (19.4-20.6) and (23-26) minutes, after injection of 4 μl cauliflower

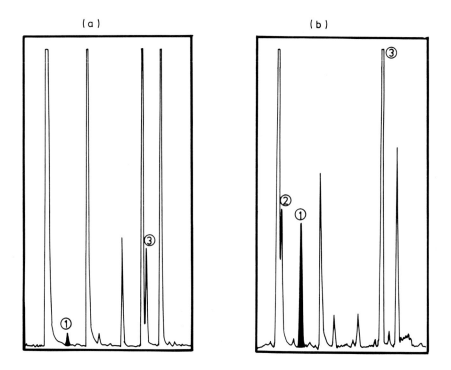

(a)   (b)

Figure 8   Total ion chromatogram of definite cuts of
            cauliflower extracts
            a) cut from 19.4-26.0 minutes on packed column,
               sample size 1 μl
            b) cut from 19.4-20.6 and 23-26 minutes, sample
               size 4 μl.

extract gave chromatogram (b). A clean mass spectrum could now be obtained and additionally a second substance (2) which was masked by a major component could be detected. At the same time it was possible to enrich component (3) and eliminate the substance eluting immediately before.

2. Enrichment of trace components with repetitive trapping.
   For the determination and identification of trace components by means of mass-spectrometric analysis clean and
   intense mass spectra are required; in these cases repetitive pre-separations can be a good approach. The cumulated
   material in the trap after several cuts is subsequently
   separated by the capillary column and can be detected with
   a mass spectrometer.

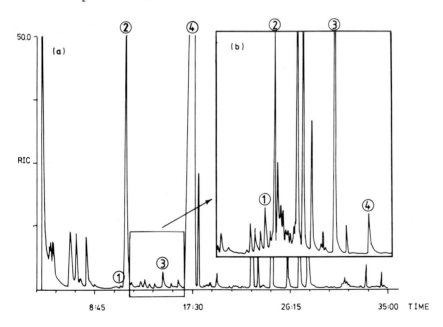

Figure 9   Comparison of TIC-chromatogram obtained with 1 µl
           sample of cauliflower extract with normal capil-
           lary (a); after repetitive trapping of substances
           eluting between 7 - 11 minutes on the packed
           column (b).

In Fig. 9 trace components have been enriched for better
identification. The chromatogram (a) was obtained with the
original sample, the chromatogram (b) was obtained after
enrichment with two successive runs of 3 microlitres each.
It can bee seen that all substances between component 2
and 4 have been enriched in comparison to chromatogram (a).

3. <u>Quantitative determination of selected compounds.</u> Different
   injections were performed in order to find out if quantification of selected compounds is possible. The results of
   these experiments are shown in Fig. 10. Chromatogram (a) was
   obtained after injection of 1 µl sample on the packed

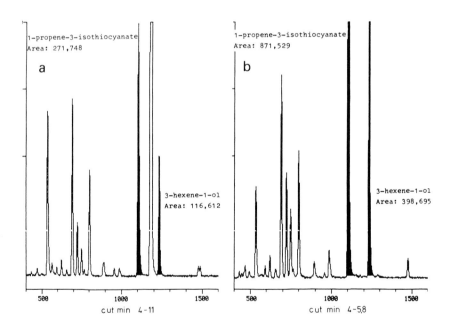

Figure 10 Comparison of chromatographic area versus sample
           size
           a) cut from 4 - 11 minutes, sample size 1 µl
           b) cut from 4 - 5.8 minutes, sample size 3 µl.

column and cutting the zone between 4 and 11 minutes. Chromatogram (b) represents a cut between 4 - 5.8 minutes after injection of 3.3 µl sample material. From a preliminary experiment it was known that 1-propene-3-isothiocyanate and 3-hexene-1-ol are eluted between 4 - 5.8 minutes on the packed column. The second cut was stopped at 5.8 minutes in order to avoid probable interference of compound (1) with 3-hexene-1-ol. The resulting peak areas deviated from the expected values about 5 %. This is a fairly good result, keeping in mind, that this cut experiment was carried out in the region of solvent tailing. The preliminary conclusion from these experiments is that quantification after trapping and desorption with this system seems to be possible.

Conclusion

Up to now intermediate trapping on adequate materials seems to be the most promising method for enrichment of trace organics. The application of this technique is a powerful tool for the analysis of gaseous samples and, in combination with multidimensional gas chromatography, for the determination of trace compounds in complex mixtures.

References

1. Ohloff, G.: Helv. Chim. Acta 65, 1785 (1982).

2. Murray, K.E.: Journal of Chromatography 135, 49 (1977).

3. Dravnieks, A., Krotoszynski, B.K., Whitfield, J., O'Donnel, A., Burgwald, T.: Environ. Sci. Technol. 5, 1220 (1971).

4. Zlatkis, A., Lichtenstein, A.A., Tishbee, A.: Chromatographia 6, 67 (1974).

5. Zlatkis, A., Liebich, H.M.: Clin Chem. 17, 592 (1971).

6. Teranishi, R., Mon, T.R., Robinson, A.B., Cary, P., Pauling, L.: Anal. Chem. 44, 18 (1972).

7. Rushneck, D.R.: J. Gas Chromatog. 3, 319 (1965).

8. Bartle, K.D., Bergstedt, L., Novotny, M., Widmark, G.: J. Chromatog. 45, 256 (1969).

9. Altshuller, A.P.: J. Gas Chromatog. 1, 6 (1963).

10. Pellizari, E.D., Carpenter Ben, H., Bunch, J.: Environ. Sci. Technol. 9, 556 (1975).

11. Ott, U., Liardon, R.: Flavour 81, 3rd Weurman Symposium, Proceedings of the International Conference, Munich, April 1981. Editor: P. Schreier; Walter de Gruyter & Co., Berlin-New York.

12. Kaiser, R.E.: Proceedings of the 8th International Symposium on Advances in Chromatography, 1973, page 215.

13. Neu, A.J., Merz, W., Panzel, H.: J. of HRC & CC 5, 382 (1982).

14. Schomburg, G., Husmann, H., Weeke, F.: J. Chromatogr. 99, 63 (1974).

15. Schomburg, G., Husmann, H., Weeke, F.: J. Chromatogr. 112, 205-217 (1975).

16. Schomburg, G., Husmann, H.: Chromatographia 8, 517 (1975).

17. Schomburg, G., Behlau, H., Dielmann, R., Weeke, F., Husmann, H.: J. Chromatogr. 142, 87 (1977).

THE ANALYSIS OF TRACE COMPONENTS USING THE TOTAL TRANSFER TECHNIQUE IN COUPLED COLUMN SYSTEMS

Manfred Oreans, Friedhelm Müller, Dietrich Leonhard and Achim Heim
Siemens AG, D-7500 Karlsruhe

Introduction

There are several possibilities to increase the performance of an existing chromatographic system. A technique which has already been described in earlier papers (1-4) is the column switching technique. Especially the "Live Column Switching Technique" as developed by Siemens in cooperation with Schomburg and Weeke shows how easy it is to achieve better results even with a capillary system. Fig. 1 shows the principle of the "Live" system for two capillaries. The modes "straight", "cut", and "backflush" are possible with this system.

What are the advantages of the "Live System"? These are the following ones:
- use of a zero-dead volume union tee for coupling of the columns,
- no valves in the flow path,
- fast switching without baseline affecting,
- current monitoring of the chromatogram after the first column and thus precise setting of cut times,
- quantitative analyses through complete sample transfer to the monitor- or main-detector,
- use of inert materials (glass, quartz, platinum) in the flow path and thus no decomposition or discrimination of the sample.

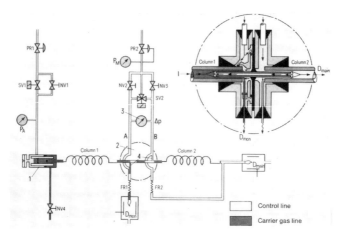

Fig. 1   Principle of the "Live Column Switching Technique"
         for two capillaries

Nevertheless, a capillary system has not only advantages. It
exhibits low sample concentration tolerance, and therefore the
sample is often splitted in the injector, i.e. as to trace
analyses in many cases the sample amount is too small to get
good results. Furthermore, extremely high solvent concentrations cause problems by covering small peaks of interest, or
because of a too difficult sample clean-up a lot of inter-
fering peaks may affect the analysis.

Another possibility, the coupling of a packed column with a
capillary one (Fig. 2) avoids some of the described problems.
In this way, the capillary can be protected against impurities
by switching only peaks of interest into the capillary. As in
this system the split point is brought from the injector to
the "live" union tee, troubles concerning discrimination etc.
can be avoided. However, the split still remains. It has to be
pointed out that at the beginning of the capillary now a trap
is needed to focus the peaks. That was the reason for us to
develop a new system, which eliminates all the mentioned pro-

blems. This system for the coupling of packed columns with capillaries - the so-called "Total Transfer System" - is described in this paper.

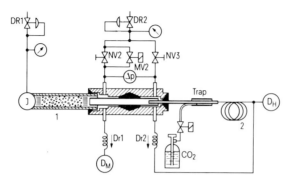

Fig. 2   Coupling of a packed (1) with a capillary column (2)

Experimental

The experiments were carried out on the twin oven gas chromato-graph SICHROMAT 2. Up to now this gas chromatograph is only suitable for the installation of the "Total Transfer System" because it is built in the modified partition wall of the SICHROMAT 2 (Fig. 3). Moreover, it is possible to set the best temperatures in both ovens for the packed column as well as for the capillary.

How does the "Total Transfer" work? Fig. 4 shows the design of the "Total Transfer" tee union. The coupling of the packed col-umn with the capillary one is performed by a glass tube which is the real trap. This connecting tube can be empty, wall-coated or filled with packing material of about 10 mm. Around the tube there is a heating wire for fast heating up the trap and thus ejecting the trapped components. The cooling of the trap tube is performed by a second glass tube designed as a mantle around the trap. The cooling medium is nitrogen, which flows in a cop-

174

per tube through liquid nitrogen in a dewar vessel. The inlet
of the cooling medium is on the capillary side to provide a
good focusing of the peaks.

Fig. 3 "Total Transfer System"

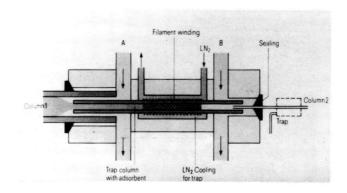

Fig. 4 Design of the "Total Transfer" tee union

Figure 5 shows the three possible switching modes of the "Total Transfer".

1. Carrier flows from the injector through the packed column
to the "Total Transfer" (TT) tee union and then through the
flow restrictor 1 to the monitor detector. This is provided by
the following: carrier comes from the pressure controller 2

MAIN FUNCTIONS OF TOTAL TRANSFER TRAP

Sample to detector 1.
Pressure at A a little bit
lower than at B.

Sample to trap column.
Pressure at A higher
than at B. Flow rate in
column 1 equal to the flow
rate in the trap column.

Injection of the sample into
column 2 and backflush of column 1.
Pressure at A a little bit higher than at B.
Flow rate in the trap column equal
to flow rate in the capillary.

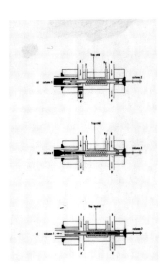

Fig. 5   Main functions of "Total Transfer Trap"

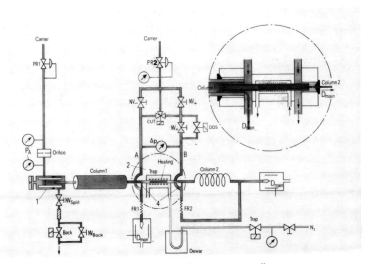

Fig. 6   Scheme of "Total Transfer System"

via the CUT-valve and the DOS-valve to the tee union and is
then distributed to the capillary column, the trap column and
the flow restrictor 2. A second flow comes via needle valve
NV- to the A-side of the tee union and from there via flow
restrictor 1 to the monitor detector. Needle valve NV- causes
a drop in pressure; i.e. compared with the B-side there is a
lower pressure at the A-side, i.e. in the trap column flows
carrier from B to A.

2. The whole carrier flow of the packed column comes into the
trap column and from there through flow restrictor 2 to the
main detector. This is provided by the carrier flow of pressure
controller 2, which comes via the CUT-valve to the A-side, and
is distributed there to flow restrictor 1 and to the trap column. The carrier flow to the B-side is restricted by needle
valve NV+, so that a big differential pressure from A-side to
B-side results.

3. Through the heated trap column less carrier gas flows than
suitable for capillary work, so that the whole sample amount
comes from the trap column to the capillary. This is provided
by carrier flow of pressure controller 2, which flows via the
CUT-valve to the A-side and is distributed there to the packed
column for backflushing it to the flow restrictor 1 and to the
trap column. Carrier flows to the B-side through the opened
DOS-valve. With needle valve NV+ the flow is restricted this
way, i.e. from the A-side to the B-side a small differential
pressure results.

Results and Discussion

The advantages of a chromatographic system like the "Total
Transfer System" are:
- Injection of relatively high concentrations into the capil-
  lary by one or several injections into the packed column and

trapping of the sample before injecting it into the capillary.

- Best protection of the capillary and thus longer durability and better separation efficiency for a longer time by cutting only interesting components into the trap.

These advantages make the "Total Transfer System" ideal for sub-trace analysis. Nevertheless, the demands for the user who works with such a system are increasing. He has to find out the following parameters for every particular analytical problem:
- trap temperature,
- ejection temperature for the trap,
- trap capacity depending on the interesting components,
- response factor of the system.

In the following, a practical example will be presented. In Fig. 7 a chromatogram of different fatty acid methylesters and hydrocarbons on a packed column is shown. The separation is insufficient, and components present in low concentrations show broad, small peaks.

Figure 8 shows the same sample on the "Total Transfer System". All components are well separated and also for substances in low concentrations good peaks resulted.

In Figure 9 a sample with only palmitic- and stearic acid methylester solved in hexane is shown. The left chromatogram shows the sample on the packed column, the chromatogram on the right side shows indisputably the advantage of the "Total Transfer System" with very distinct peaks of the two esters.

The analytical data for this sample:
    Gas chromatograph: SICHROMAT 2
    Sample: 200 µg palmitic acid methylester and stearic acid
            methylester in hexane.

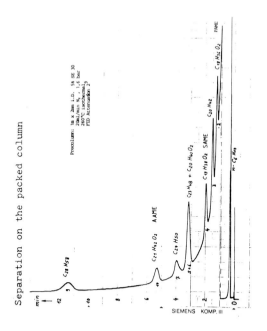

Fig. 7   Analysis of fatty acid methylesters (palmitic acid
         methylester PAME; stearic acid methylester, SEAME;
         arachidonic acid methylester, AAME; behenic acid
         methylester, BAME; and some hydrocarbons on a 5%
         SE 30 precolumn, 1 m, 2 mm i.d.; 260°C isothermal;
         20 ml/min N_2 at 1.6 bar; FID; attn. 2^9

Oven 1

Column: 2 m x 1/4" with 2 mm i.d. glass,
        5% SE 30 on Chromosorb G
Carrier: 20 ml/min N_2
Temperature: 200°C isothermal

Trap column

Effective trap: glass tube with 1.8 mm i.d., packed for
                10 mm with 5% SE 30 on Chromosorb G
Carrier: 30 ml/min for trapping
         2,5 ml/min for ejecting

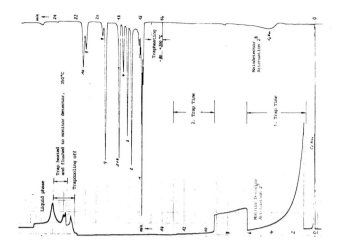

Fig. 8 The sample (cp. Fig. 7) analyzed on the "Total Transfer System"

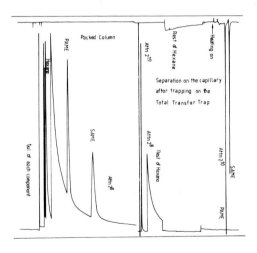

Fig. 9 Analysis of 200 µg palmitic acid methylester and stearic
acid methylester in hexane (left: sample on the packed
column; right: sample in the "Total Transfer System")

Oven 2

Column: glass capillary 20 m x 0.28 mm i.d. OV1
Carrier: 2.8 ml/min N$_2$
Temperature: 140°C, 180°C, 5°C/min

Injections with different sample volumes and different concentrations of the two esters were carried out. It was stated
that the system was absolutely linear for different sample
volumes and that there was only a 5% deviation of linearity
for a concentration range from 2 μg to 200 μg. It is evident
that the detectivity has increased by a factor of about five
through the distinct peaks of the capillary. The reproducibility for several injections was better than 99.5%.

At least, three chromatograms of cauliflower flavour extract
are represented. Figure 10 shows an extract on a packed column
with only few peaks. Figures 11 and 12 show two different heartcuts of the packed column after trapping and ejecting of the
trap into the capillary.

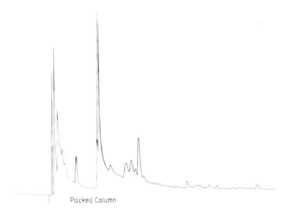

Packed Column

Fig. 10

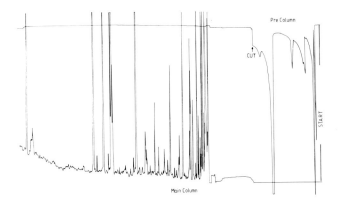

Fig. 11

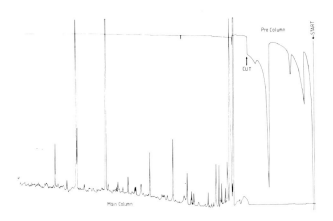

Fig. 12

Fig. 10-12   GC analysis of a cauliflower flavour extract on a
packed column (Fig. 10), and after two different
heartcuts of the packed column after trapping and
ejecting of the trap into the capillary (Fig. 11
and 12; from the top)

Acknowledgements

We thank Prof. Dr. Drawert and Dr. Nitz of the Technical University of Munich for preparing these chromatograms in their laboratories.

References

1. Müller, F., Oréans, M., Weeke, F.,: Labor-Praxis 5, 462-472 (1982).

2. Müller, F., Oréans, M.: Siemens Application Note No. 282.

3. Schomburg, G., Weeke, F., Müller, F., Oréans, M.: Chromatographia 16, 87-91 (1982).

4. Schomburg, G., Bastian, E., Behlau, H., Weeke, F., Oréans, M., Müller, F.: Proceedings of the Fifth International Symposium on Capillary Chromatography, Riva del Garda, 381-395 (1983).

CAPILLARY GC-FTIR ANALYSIS OF VOLATILES: HRGC-FTIR

Werner Herres

Bruker Analytische Meßtechnik GmbH
Wikingerstr. 13, D-7500 Karlsruhe 21

1. Introduction

For quite a long time chemists have used infrared spectroscopy
(IR) as a powerful tool in analysis of molecular structure and
in monitoring changes during preparative work. The informative
content of an IR-spectrum is very high, each molecule possessing its own characteristic IR-"fingerprint". The capability
to give unambiguous identification of a compound makes IR
spectroscopy well suited to be used as a mean of identification in conjunction with gas chromatography (GC).

In 1967, first approaches were made to use a Fourier transform
IR (FTIR) as a detector coupled to a GC (1). Already the early
GC-IR experiments demonstrated the feasibility of the technique,
although high capacity packed columns and sometimes even
stopped flow mode had to be used (for review of early works
see (2)). Since then spectrometer hardware has evolved considerably, computers got more powerful for less cost, and increasing interest in the technique brought us some theoretical
insight, some empirical optimization proposals, and quite a
flexible software. The coupling of FTIR with packed column GC
achieved a high degree of optimalization. Nowadays, high sensitivity FTIR instrumentation is available fast enough to acquire several full mid-IR spectra even over a narrow capillary
GC-peak. Concerning gas chromatography, recent developments in
this field brought us exciting new column types, injection
techniques, and column switching devices. This paper will focus

on the use of capillary columns for the hyphenated technique
GC-FTIR: HRGC-FTIR.

2. Some Basic Considerations on the Experimental Set-up

In a typical GC-IR measurement the effluent from the column is
passed down a heated transfer line and continuously through a
specially heated IR measuring cell, the "light-pipe". The experimental set-up used for the measurements reported in this
paper is shown schematically in Fig. 1. For most HRGC-FTIR runs
fused silica capillary columns were used. In any case the
fused silica column or a fused silica transfer line ended up
directly at the entrance of the light pipe. Thus the separation
performance of HRGC was maintained right into the IR measuring
cell.

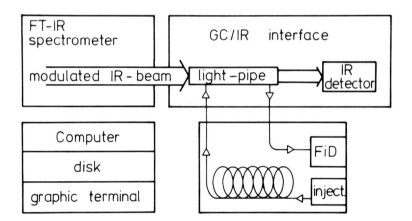

Fig. 1   Scheme of the HRGC-FTIR system used: FTIR optics,
          GC-IR interface, GC, data system with fast disk for
          data acquisition and spectral libraries

At the light pipe entrance accelerating gas was added parallel
with the effluent to maintain linear gas velocity through the
whole GC-IR interface. At the end of the light pipe a fused
silica transfer line was used to guide part of the effluent
back into the GC and to the flame ionization detector (FID).
Thus infrared spectra, infrared chromatograms and a conventional GC-FID trace for comparison were generated simultaneously during each single GC-IR run.

It is worthwhile to mention here that infrared detection is non-
destructive. So the effluent having passed down the light
pipe is totally available for other detection techniques, fraction collection, etc. We will not discuss principles of Fourier
transform IR spectroscopy. These can be looked up elsewhere,
e.g. in (3,4). The important point to know is that the FT technique gives us advantages concerning speed, sensitivity and
spectral accuracy compared to conventional IR. For this work
the IFS 85 FTIR spectrometer was used at a speed to acquire 10
to 12 complete mid-IR spectra within 1 second at a spectral resolution of 8 cm^{-1}.

The relation between the different information generated during
a GC-IR run and the transformation of raw data to a final IR
spectrum is shown in Fig. 2. The modulated infrared beam leaves
the interferometer, is focussed into the light pipe, exits the
light pipe and produces at the IR detector the "interferogram"
as raw signal (Fig. 2A). During the whole GC-IR run typically
10 interferograms are acquired every second. From the interferograms a total IR chromatogram is calculated in real time
(Fig. 2B). If we look at the situation in the light pipe when
no compound is eluting, we get a single beam background spectrum showing characteristics of optics, cell and detector.
Carrier gases He, H_2, N_2 are transparent for the IR beam (Fig.
2C). If a compound passes down the light pipe we get a single
beam spectrum, which should show some absorptions additional
to those in the background (Fig. 2D).

186

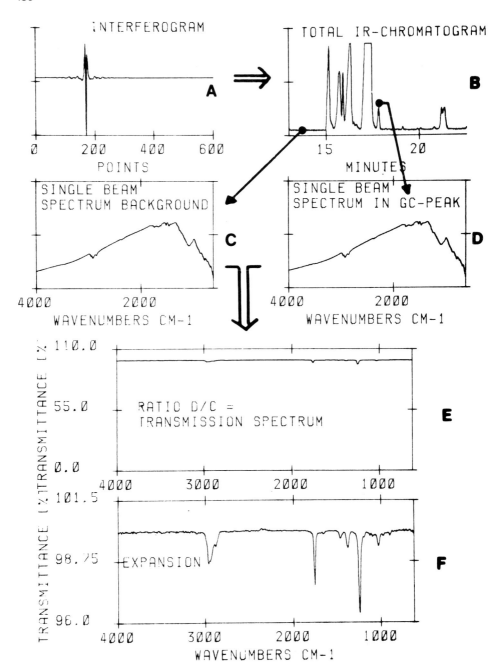

Fig. 2   Relation between raw and refined data of a GC-run

Using the background single beam as reference or I_o we calculate as usual an IR transmission spectrum by rationing I/I_o, i.e. D/C (Fig. 2E). But in HRGC-FTIR concentration of compounds are low resulting in weak absorptions. We must apply large expansion to reveal spectral features (Fig. 2F). Next step could be a library search for compound identification.

At this point it is important to realize that in HRGC-FTIR we have to deal with weak absorptions on a huge background. This is a worst case in signal processing if small changes riding on large a.c. signals must be processed and detected without distortion and loss in linearity.

The data concerning the equipment used is summarized in Table 1.

Table 1   HRGC-FTIR equipment used

Spectroscopy

Spectrometer:	BRUKER IFS 85
Light pipe:	BRUKER type 1 (360 x 1.5 mm i.d.)
IR detector:	Hg-Cd-Te (4800 - 600 cm^{-1})
Time resolution:	10-12 spectra per second at 8 cm^{-1}; in most runs 4 to 6 interferograms were added up in real time resulting in effective time slices of 0.3 to 0.5 seconds

Gas chromatography:

Gas chromatographs:	1) Carlo Erba HRGC 4160
	2) Siemens Sichromat 2

Columns:	fused silica and glass capillary co-lumns of different polarity (see ex-amples for column types used and other details)
Carrier:	He (also used as accelerating gas)
Injection techniques:	split and on-column

3. Optimization Considerations for Light Pipe, IR-Detector and Spectral Resolution

In practice the combination of FTIR with HRGC builds up many parameter optimization problems. Already the base FTIR-spectrometer should have the appropriate sophisticated control and signal processing electronics. The computer should have sufficient number crunching capability, and data storage pe-riphery should be fast enough for HRGC-FTIR data streams and large enough to hold spectral libraries as well.

3.1 Light Pipe

The design of the light pipe has to meet quite a number of re-quirements both from spectroscopy and chromatography.

As shown in Fig. 3 the IR beam is focussed into the light pipe which is essentially a gold-coated glass tube with IR trans-parent windows and transfer lines mounted at the ends. The IR beam undergoes multiple reflections by the gold coating, leaves the cell and is focussed onto the element of a liquid nitrogen cooled Hg-Cd-Te (MCT) detector. Obviously the volume of the light pipe is optimal when matched to the volume of the GC-peaks. Therefore, the volume of light pipes dedicated to HRGC-FTIR work is less than 0.1 ml (5). If packed column work has

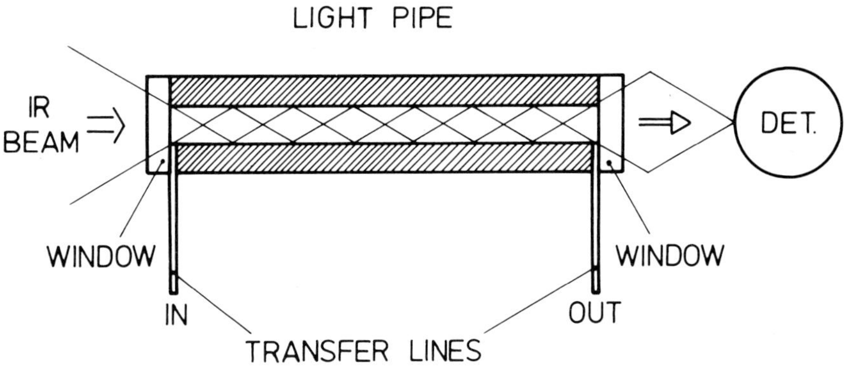

LIGHT PIPE

IR
BEAM ⟹

DET.

WINDOW   WINDOW

IN   OUT

TRANSFER LINES

Fig. 3 Light pipe (schematically)

to be done as well, a 0.6 ml volume light pipe can be chosen
for capillary work, if accelerating gas is added to the effluent at the light pipe entrance to maintain linear gas velocity.
Griffiths (6) has discussed length/diameter optimization from
the point of light throughput. Chromatography requires that
the light pipe must be chemically inert, must provide laminar
flow and should avoid dead volumes at the flanges.

3.2 IR Detector

There is a range of MCT detectors available for GC-IR work
differing in spectral bandwidth. Fig. 4 shows that broad band
detectors are less sensitive than narrow band detectors.

If utmost sensitivity is needed, a narrow band MCT is the
right choice. In practice the spectral range from 850 cm^{-1}
down to 600 cm^{-1} is quite important and particularly useful
for the identification of substitution patterns of aromatic
compounds. Therefore, a detector bandwidth 4800 to 600 cm^{-1} is
an acceptable compromise between sensitivity and useful spectral range.

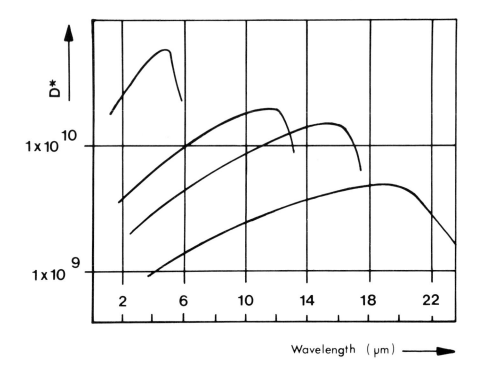

Fig. 4 Typical detectivity curves for different MCT detectors

3.3 Spectral Resolution

IR spectroscopy during a GC-IR experiment is done at normal
pressure on compounds in the gas phase. Under GC-IR conditions
we will not find the narrow bandshapes as known from low pressure gas spectra of small molecules. Experience shows that for
most compounds under investigation a spectral resolution of 8
or 4 cm^{-1} can be used without loss of important spectral information. Throughout this work a resolution of 8 cm^{-1} was used,
and one level of zero filling applied before Fourier transformation resulting in one spectral point every 4 cm^{-1}.

4. Gas Phase versus Condensed Phase Spectra

The spectrum of an isolated molecule in the gas phase can be
quite different from the condensed phase spectrum. This holds
particularly for molecules with strong intermolecular interactions such as hydrogen bonding. As an example Fig. 5A shows
the transmission spectrum of methanol (liquid) compared to its
vapor phase spectrum from a GC-IR run. If molecules become
larger, gas phase and condensed phase spectra become more
similar (see Fig. 5B). In most cases a slight shift of band
positions to higher wave numbers is observed in gas phase spectra due to the lack of a reaction field. For some molecules the
elevated temperature can enable internal rotations which are
hindered at normal temperature.

5. Sensitivity of HRGC-FTIR and GC Optimization

During a HRGC-FTIR run the IR data are acquired "on-the-flow".
Thus the time available to acquire the spectrum of an eluting
compound correlates with its GC peak width. The amount of compound needed to obtain a sufficiently structured IR spectrum
is 5 to 20 ng for "good" IR absorbers. This is not only true
for "bright sunshine" samples like isobutylmethacrylate. Fig.
6 shows a spectrum as obtained during a real-world HRGC-FTIR
run. The sample was of environmental relevancy. A 30 m x 0.32
mm i.d. fused silica column DB-1 was used for separation after
on-column injection. Note that transmission scale going from
98.7% to 99.3% = 0.6%! Although signal to noise (S/N) is approximately only 4 to 10, this spectrum is sufficiently structured
to give positive identification of the unknown compound by
library search: m-cresol.

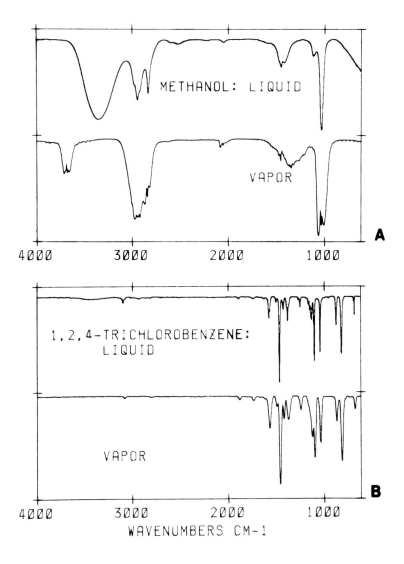

Fig. 5 A) Transmission spectrum of methanol (liquid) and
            methanol in gas phase
        B) Transmission spectrum of 1,2,4-trichlorobenzene as
            neat liquid and in gas phase

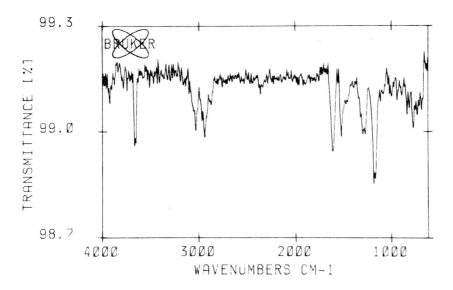

Fig. 6   Spectrum of a weak chromatographic peak from HRGC-FTIR
          run on an environmental sample. Identification by
          library search: m-cresol

This situation is different for compounds where the useful information is among the weaker spectral features. The spectra
of cis- and trans-decahydronaphthalene (Fig. 7, A/B) may demonstrate this. The strong absorptions near 3000 cm^{-1} do not
help for identification. To reveal the weaker spectral features
here and in comparable cases with a S/N of 4 to 10 the amount
of compound injected should be in the 50 to 120 ng range.

194

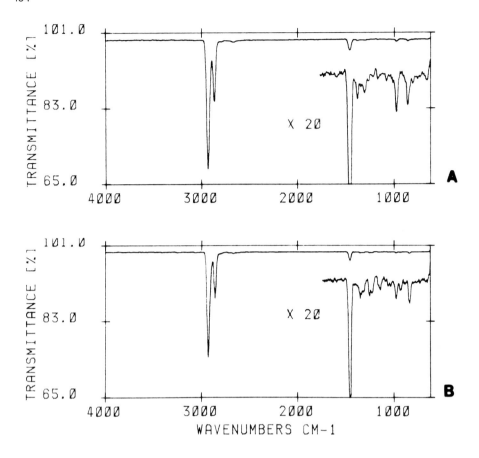

Fig. 7   GC-IR spectra of cis-decahydronaphthalene (A)
         and trans-decahydronphthalene (B)

Now we can discuss the question: What type of capillary column
should be selected for HRGC-FTIR work? Fig. 8 shows for common
WCOT column diameters schematically the relation between diameter, capacity and efficiency. A scale for IR spectral signal
to noise is added. From Fig. 8 it is immediately clear that
modern thick film columns offer the best compromise between
concentration requirements and separation efficiency. If available, bonded phases are preferable. The dynamic range of com-

pound concentration for HRGC-FTIR analysis is determined by
column capacity at the upper side and by IR S/N at the lower
side. Application of recently developed devices for trapping
and column switching can extend the usable dynamic range considerably. The GC itself should be equipped with an on-column
injector additional to a good split/splitless injector.

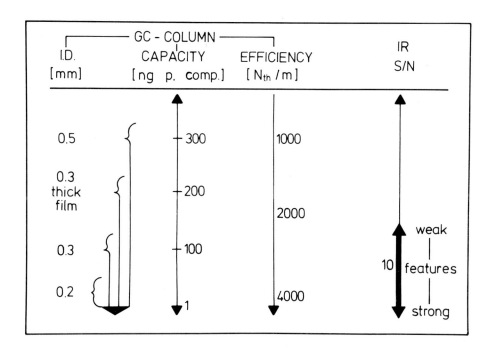

Fig. 8   Interrelation between GC column characteristics and
          IR spectral signal to noise

6. IR-Chromatograms, Quantitative Information

There are several ways to construct or reconstruct chromatograms from IR data. A very elegant method to calculate a total
IR chromatogram is the Gram-Schmidt vector projection technique. This technique is described in detail in (7). It works
in the interferogram domain, is fast and gives chromatograms

with fairly good S/N. Briefly, it takes during a GC-IR measurement a part of each just acquired interferogram and projects
it into a multidimensional reference space which was built up
when only carrier gas was passing down the light pipe. Because
- simplified - discrete spectral information is distributed
over the entire interferogram, a total IR chromatogram is generated.

In 1982, first evaluations of the quantitative capabilities of
the Gram-Schmidt technique were made (8). We have checked its
quantitative behaviour running a HRGC-FTIR series of benzene
diluted in hexane covering the concentration range 0.025 to
1 Vol.%. Using split injection the effective amount of benzene
was 20 to 1000 ng.

From Fig. 9 we see that the integrated Gram-Schmidt intensity
of the benzene peak correlates well with the injected amount

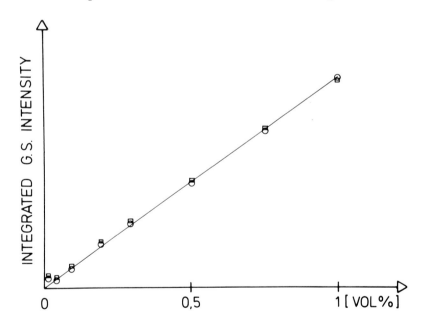

Fig. 9   Integrated Gram-Schmidt intensity for benzene in hexane
         (two independent runs)

of sample over a wide concentration range. The reason for deviation from linearity at the very low concentration is not
clear at the moment.

A different possibility to calculate IR chromatograms is the
construction of IR group specific or IR window chromatograms.
Here the change in transmission during a GC/IR run is monitored
in discrete spectral regions. This can be done working on transformed spectra or doing a "one-point" Fourier transform. If IR
window chromatograms are to be calculated in real-time during
a HRGC-FTIR run, an array of FFT processor should be added to
the computing system.

As an example for the discussed types of IR chromatograms Fig.
10 shows a HRGC-FTIR run on a mixture of methanol, isopropanol,
benzene, dimethylformamide, cis/trans-decahydronaphthalene and
2,6-dimethylaniline in CH_2Cl_2 as solvent. The trace at the top
is the reconstructed chromatogram at 3650 cm^{-1} exhibiting only
peaks for the alcohols. Window chromatogram at 1720 cm^{-1} shows
DMF peak only. Note that none of the windows is right for benzene.

7. Application Examples of HRGC-FTIR

7.1 Petrochemistry

The interest in detailed analysis of crude oils and oil distillates is great and still growing. This is for economical reasons, but for eventual environmental hazards from some components as well. The sample used for the HRGC-FTIR measurement
shown in Figs. 11 and 12 was a crude oil distillate. A non-
polar fused silica capillary column 30 m x 0.32 mm I.D. was
used for separation. The GC-oven was programmed for 2 min initial hold at 60°C, from 60° to 100° at 5°/min, and at 12°/min

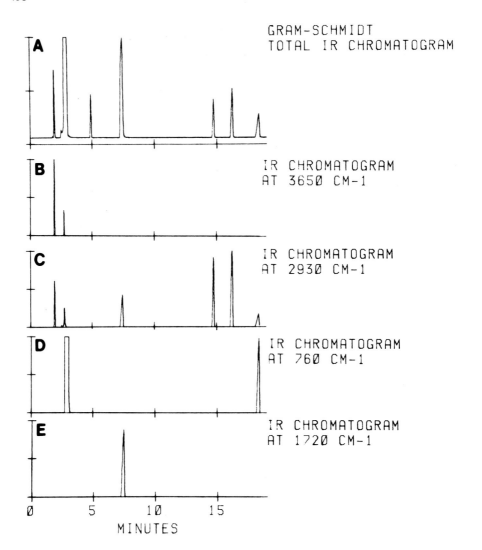

Fig. 10   Gram-Schmidt and spectral window chromatograms of test mixture

up to 200°C. Injection port, transfer line and light pipe were
at 240°, 230° and 220°C, respectively. For comparison, both
chromatograms, FID and Gram-Schmidt, are assembled together in
Fig. 11. The chromatograms are real time traces from the same
HRGC-FTIR run having the FID mounted behind the entire GC-IR
interface.

As typical for oil distillates the majority of peaks results
from aliphatic hydrocarbon compounds. But there is a number of
aromatic components as well, e.g. the spectra extracted from
the peaks at 9.25 min, 10.32 min and 16.98 min retention time
are plotted in Fig. 12. By library search the compounds could
be clearly identified to be ethylbenzene (9.25 min). o-xylene
(10.32 min) and naphthalene (16.98 min).

7.2 Analysis of Solvents and Solvent Mixtures
    Vapor Phase Libraries

The application of HRGC-FTIR to the analysis of impurities or
additives in lab grade solvents is straightforward. If on-
column injection technique is used, sensitivity goes down to
some ppm.
The situation becomes more complex if industrial solvents and
solvent mixtures are to be analysed. In many industries solvents are purified after use by distillation and other processes. They must be controlled before recycling for quality
and examined for possible health hazards. Quality control is
in most cases done by gas chromatography alone. The "identification" of unknown cancerous or other hazardous components by
looking up retention time tables bears many problems. HRGC-FTIR
analysis gives both detection and identification from a single
measurement.

The sample we will deal with in this chapter is an industrial
solvent mixture. In addition this mixture was spiked with a

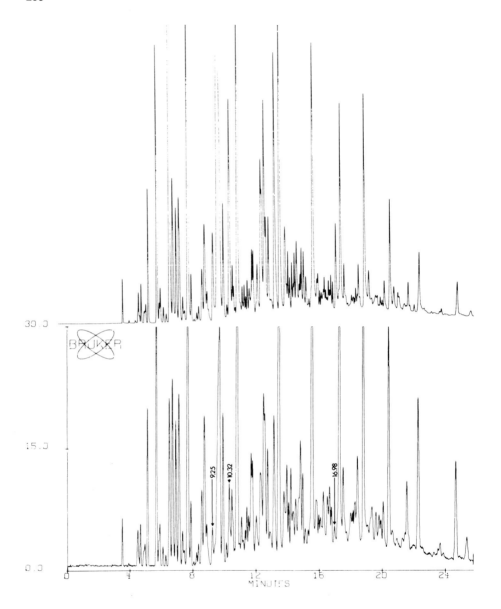

Fig. 11   HRGC-FTIR measurement of a crude oil distillate.
TOP: GC-FID detector trace, FID mounted behind the
entire GC-IR interface.
BOTTOM: Real-time Gram-Schmidt total IR chromatogram
generated during the same GC-IR run

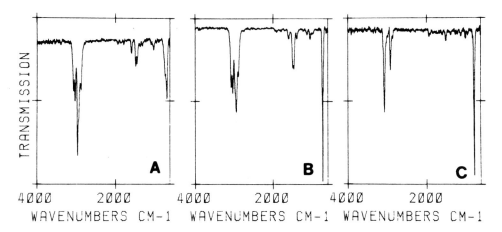

Fig. 12   Spectra of some aromatic compounds in the mostly ali-
           phatic crude oil distillate. A) ethylbenzene (9.25
           min), B) o-xylene (10.32 min), C) naphthalene (16.98
           min)

number of pollutants. A SE 30 type fused silica capillary column was used for separation (30 m x 0.32 mm I.D.). Temperature
programming was 2 min initial hold at 50°C, 10°C/min to 90°C,
4°/min to 110°C and 6°/min to 200°C. The chromatogram of the
HRGC-FTIR run is shown in Fig. 13. We used this industrial
solvent mixture to test the available IR vapor phase libraries
for their quality. A spectral library search software and a
reliable spectral library is important for efficient GC-IR
work. There are currently two IR vapor phase libraries available. The larger library covering 7800 compounds and still ex-
panding is commercially marketed by Sadtler (9). A smaller
library of approximately 2300 spectra was financed by the US
government (so-called "EPA library").

Both libraries were adapted for on-line use on the mini computer of the FTIR spectrometer (10). At the retention times
marked in the chromatogram (Fig. 13) spectra were extracted
and searches were performed in both libraries. The 10 best

matches from each search report (the "hit list") were examined
for similarity by visual inspection and superposition on the
graphic screen.

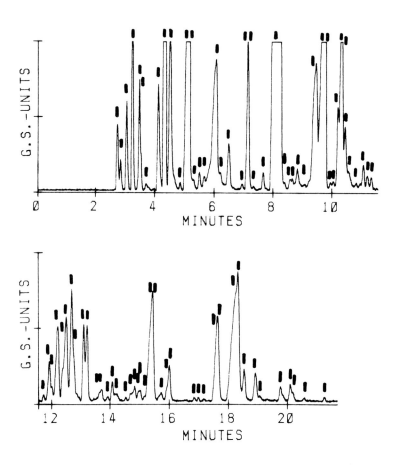

Fig. 13 Gram-Schmidt chromatogram of HRGC-FTIR measurement
of industrial solvent mixture

In Table 2 the results are summarized. All the major peaks and
most of the lower concentrated components were identified. Even
for the spectra without perfect identification the "hit list"
was of great value to classify these spectra according to certain functionalities.

Table 2
HRGC-FTIR Measurement of Industrial Solvent Mixture
Compound Identifications by Library Search

Retention Time (min)	Vapor phase library Sadtler	EPA
2.75	Water	+
2.86	Methanol	+
3.06	Ethanol	+
3.25 front	Aceton	+
3.25 back	Isopropylalcohol	+
3.49	Acetic acid, methylester	+
3.56	Methane, dichloro	+
3.73	1-Propanol	+
4.13	2-Butanone	
4.35	Acetic acid, ethylester	+
4.53	Isobutylalcohol	+
4.86	Ethane, 1,1,1-trichloro-	+
5.13 main	Butanol	+
5.13 back	-ol	
5.51	Formic acid, isobutylester	+
5.67	1-Butanol-, 3-methyl-	+
5.83	-ol, branched	
6.06	Ethanol, 2-ethoxy	
6.50	2-Pentanone, 4-methyl	+
6.96	Alkane	
7.14 front	Benzene, methyl	+
7.14 back	Acetic,acid, isobutylester	
7.34	-ester	
7.67	Alkane	
8.15	Acetic acid, butylester	+
8.41	Alkane, branched	
8.58	Acetic acid, 2-oxyester	+
8.66	Alkane, cyclic	

8.83		-ol, Ketone	
9.08		Ketone, branched	
9.47		Benzene, ethyl	+
9.76		m-Xylene	+
9.76	back	p-Xylene	+
9.96		Alkene	
10.06		Alkane	
10.21		Benzene, pentyl-	+
10.33	main	o-Xylene	+
10.33	back	Acetic acid, 2-oxyester	+
10.46		Cellosolve, butyl	+
10.61		Alkene	
10.83		-	
11.08		Benzene, isopropyl	+
11.21		Alkane	
11.35		Alkane, branched	
11.73		Alkane	
11.93		Benzene, propyl-	
12.01		Alkane	
12.20		Benzene, 1-ethyl- 3-methyl	
12.38		Benzene, 1,3,5-trimethyl	+
12.50		Phenol	+
12.68		Tartaricacid, dibutylester	+
12.77		Toluene, o-ethyl or Benzene, o-diethyl	
13.10		Alkene	
13.22		Benzene, 1,2,4-trimethyl	+
13.65		Alkene	
13.71		Alkane	
13.93		Alkene	
14.10		Benzene, 1,2,3-trimethyl	+
14.20		Alkene	
14.53		Indene, 2,3-dihydro	
14.68		o-Cresol	

14.82	Benzene, 1-ethyl-3-methyl or m-diethyl	
14.96	Alkene	
15.02	Cumene, 3,5-diemethyl-	+
15.16	Alkane	
15.40	p-Cresol	+
15.40 back	m-Cresol	+
15.73	Acetic acid, 2-oxyester	
15.90	Acetic acid, ester	
16.00	Acetic acid, oxyester	
16.83	Alkene or aromatic	
16.97	Benzene, 1,2,4,5-tetramethyl	+
17.17	Phenol, 2-ethyl	+
17.63	Cymenol or Xylenol	
17.63 back	Phenol, 2,5-dimethyl	
18.22	Phenol, 4-ethyl-	+
18.32	Phenol, 3,5-dimethyl-	
18.54	3-Xylenol	
18.92	4-Xylenol	
19.08	Phenolic comp.	
19.78	Phenol, p-sec-butyl-	+
20.10	p-Cymenol	
20.20	Phenolic comp.	
20.60	p-Phenolic comp.	
21.26	Ester	

7.3 Flavor and Fragrances

HRGC-FTIR was recently applied successfully to the investigation of tropical fruit volatiles (11-13). The aroma composition of fruits is mainly determined by nonpolar hydrocarbons,

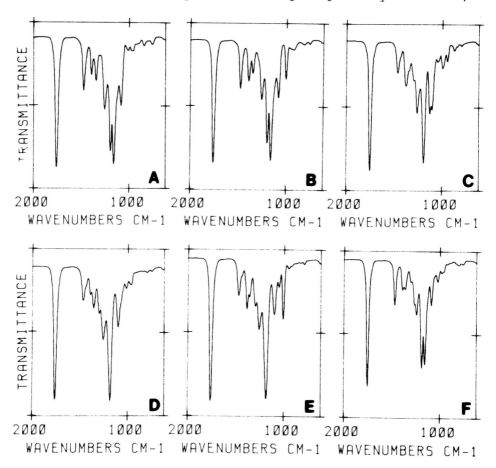

Fig. 14 HRGC-FTIR spectra of C_4-C_4 esters: A) n-butyl isobutanoate, B) 2-butyl isobutanoate, C) 2-butyl n-
butanoate, D) n-butyl butanoate, E) isobutyl n-butano-
ate, F) isobutyl isobutanoate

esters and alcohols. From the various esters creating a "fruity" smell, the different butanoates are rather difficult to identify from their retention indices and mass spectra. The IR vapor phase spectra of the C_4-C_4 esters show clear differences in the fingerprint region allowing positive identification of the different isomers (Fig. 14 A-F).

The capability to discriminate between different isomers makes HRGC-FTIR well suited for the analysis of essential oils. The spectra of neral and geranial (Fig. 15A,B) for example show significant differences. Even in the case of menthone/iso-menthone there are discriminating spectral features in the fingerprint region (Fig. 16A,B).

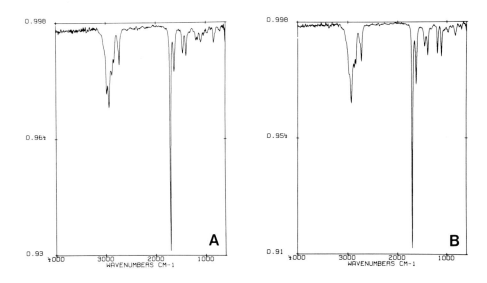

Fig. 15   HRGC-FTIR spectra of A) neral and B) geranial from
          an analysis of cold pressed lemon oil

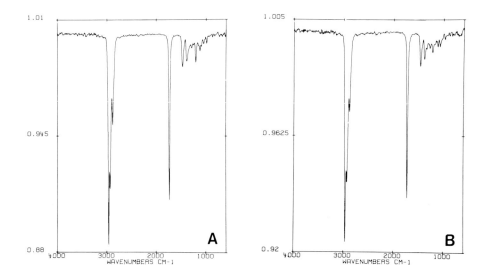

Fig. 16   HRGC-FTIR spectra of A) menthone and B) iso-menthone
          from an analysis of peppermint oil

But it is obvious from Fig. 16A,B that in this case discrimination needs the weaker spectral features. This situation is
typical for terpene and sesquiterpene compounds. As an example
for the latter the spectra of bergamotene and ß-bisabolene are
shown in Fig. 17A,B.

Another point to realize when doing HRGC-FTIR analysis of
essential oils is the high dynamic range of compound concentration in many of these oils. If we do a HRGC separation of a
peppermint oil for example (Fig. 18), quite optimal GC-conditions result in concentrations of the minor components well
below the IR detection limit.

Therefore, the IR Gram-Schmidt chromatogram of this separation
(Fig.19,top) only reveals the stronger peaks.To get IR spectra

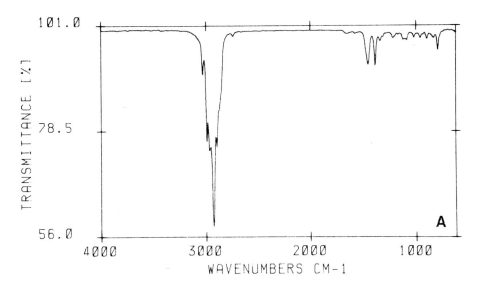

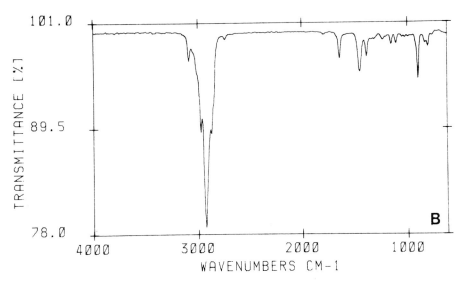

Fig. 17   HRGC-FTIR spectra of A) bergamotene and B) ß-bisabo-
          lene from HRGC-FTIR analysis of a sesquiterpene mix-
          ture

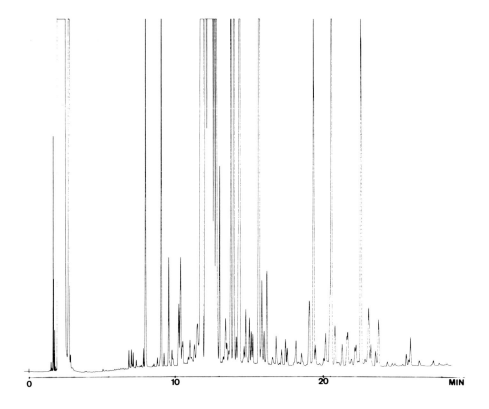

Fig. 18   HRGC separation of peppermint oil (FID trace):
on-column injection of 1μl of 0.5% solution in hexane
(30 m x 0.3 mm I.D. fused silica DB-1 column)

of the minor components we have to raise their amount into the
5 to 50 ng range. This was done for the measurement in Fig. 19
(bottom) using the same chromatographic conditions as outlined
in Fig.18. The column was heavily overloaded, but separation in
second part of the chromatogram was maintained and well structured spectra from the minor peaks could be extracted. By
changing temperature programming the early part of the chromatogram was optimized concerning compound concentration vs.
chromatographic separation. There is a very elegant way to deal

BRUKER GC-FTIR : PEPPERMINT OIL

GRAM-SCHMIDT TRACES

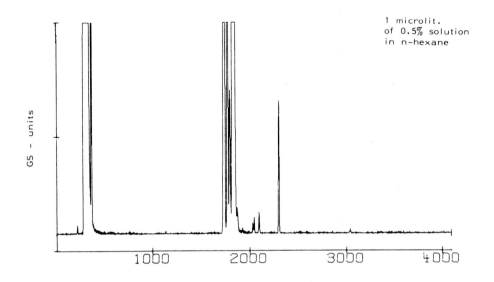

1 microlit.
of 0.5% solution
in n-hexane

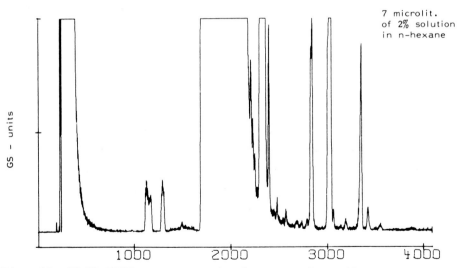

7 microlit.
of 2% solution
in n-hexane

Fig. 19   HRGC-FTIR measurement of peppermint oil
(Gram-Schmidt traces, time-axis in scan sets.
See text for details)

with similar high dynamic concentration range samples using
column switching techniques. These will be discussed in chapter 8.

7.4 Other Applications

In Fig. 6, a spectrum from a HRGC-FTIR separation of an environmental sample was shown. In fact pollution control is an
important field of application for HRGC-FTIR. Many of the pri-
ority pollutants give highly significant IR vapor phase spectra
at a low detection limit. Among other fields becoming important
are head space analysis and the analysis of volatiles after degradation of polymers by thermal treatment or pyrolysis.

8. Column Switching Techniques

The modern capillary columns with bonded phases can be used to
couple serially columns of different polarity. Even if a
straight coupling without flow switching is applied the combination of different polarities can shorten analysis time for
routine GC samples considerably.

We tested a more sophisticated coupling technique for its potential in combination with FTIR. The experimental set-up is
shown schematically in Fig. 20: in a dual oven GC one capillary
column was mounted in each oven. Column coupling was done via
a "life" column switching device (14). This switching device
was used in straight transfer mode, backflush mode and "cut
mode". The application of backflush technique in combination
with a precolumn/capillary column set-up is straightforward
and can be done with much simpler devices. The main interest
was in the feasibility of the "cut mode". In this mode the
effluent from column no. 1 is switched during a GC run at arbitrary or programmable retention times to the second column or

to a monitor FID.

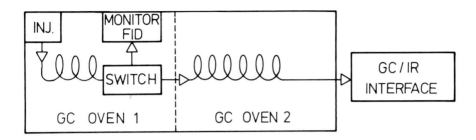

Fig. 20   Scheme of dual oven GC with "life" column switching

This technique in combination with FTIR was successfully
applied to some problems in oil distillate analysis using a
50 m x 0.32 mm I.D. nonpolar glass capillary column in the GC
oven no. 1 and a 60 m x 0.32 mm I.D. glass capillary column of
high polarity in oven no. 2 (15).

It could be expected that the cut mode in combination with a
higher capacity column in oven no. 1 and a highly efficient
column in oven no. 2 should extend the dynamic range of compound concentrations for HRGC-FTIR analyses. Therefore, we
coupled a 30 m x 0.32 mm I.D. thick film fused silica DB-1
column in oven no. 1 with a 20 m x 0.32 I.D. Carbowax 20 type
column (CP ^{tm}57CB) with a 0.2 micron film in oven no. 2. We repeated the HRGC-FTIR analysis of peppermint oil (Fig. 18) using
a concentration similar to that used for Fig. 19B. During the
first 13 minutes of the separation the effluent from column no.
1 was switched to the monitor FID. After the bundle of major
components the effluent was switched to the second column. The
second column ended as usual at the entrance of the light pipe.
The Gram-Schmidt chromatogram of this separation demonstrates
the feasibility of the coupling (Fig. 21). There is no interference from the major components.

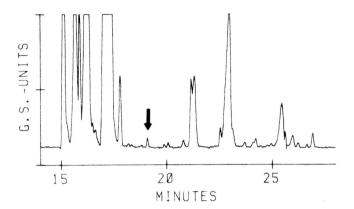

Fig. 21 HRGC-FTIR analysis of peppermint oil using cut mode
         (Gram-Schmidt chromatogram)

Clean spectra were extracted from the minor peaks. As an example the spectrum of pulegone is shown in Fig. 22. The corresponding GC peak is marked in Fig. 21.

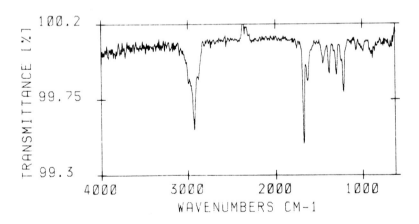

Fig. 22 HRGC-FTIR spectrum of pulegone from analysis of pepper-
         mint oil

Another example to demonstrate the applicability of column switching deals with perfume analysis. The experimental set-up was as before. The sample was an artificial perfume composition which was kindly supplied by Prof. Bruns, Thera Chemie, Krefeld, FRG. In Fig. 23A the separation on column no. 1 alone is shown. In cut mode during the two marked time periods only the effluent was switched to the second column (Fig. 23B). The spectrum of α-terpineol (Fig. 24) was acquired at the retention time marked in Fig. 23B.

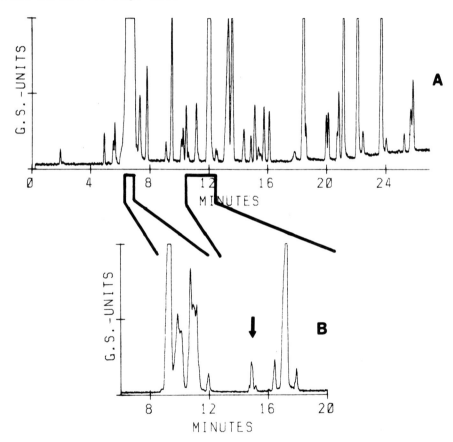

Fig. 23   HRGC-FTIR measurement of perfume composition:
   A) Gram-Schmidt chromatogram of separation on 30 m x
      0.32 mm I.D. fused silica DB-1 alone
   B) Gram-Schmidt chromatogram of separation with column
      switching applied

216

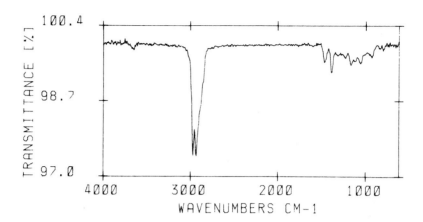

Fig. 24   HRGC-FTIR spectrum of α-terpineol from analysis of
          perfume composition

Acknowledgement

Some flavour substances in this work were kindly provided by
Prof. Dr. K.H. Kubeczka and Prof. Dr. P. Schreier, both at the
University of Würzburg.

References

1. Low, M.J.D., Freeman, S.K.: Anal. Chem. 39, 194 (1967).

2. Erickson, M.D., Appl. Spectrosc. Rev. 15, 261 (1979).

3. Griffiths, P.R.: Chemical Infrared Fourier Transform Spec-
    troscopy, Wiley & Sons, New York 1975.

4. Ferraro, J.R., Basile, L.J., Eds.: Fourier Transform Infra-
    red Spectroscopy, Applications to Chemical Systems, Vol. 1
    (1978), Vol. 2 (1979), Vol. 3 (1982), Academic Press, New
    York.

5. Bruker dedicated HRGC-FTIR interface.

6. Griffiths, P.R.: Gas Chromatography and Fourier Transform
    Infrared, 143-168 in Ref. 4, Vol. 1.

7. Hanna, D.A., Hangac, G., Hohne, B.A., Small, G.W., Wieboldt, R.C., Isenhour, T.L., J. Chromatogr. Sci 17, 423 (1979).

8. Sparks, D.T., Lam, R.B., Isenhour, T.L., Quantitative GC/FTIR Using Integrated Gram-Schmidt Reconstruction Intensities, presented at Pittsburgh Conf., 1982.

9. Sadtler Vapor Phase Library, Sadtler Research Labs., USA.

10. BIRSY - Bruker Infrared Spectral Search System, Bruker, W. Germany.

11. Herres, W., Idstein, H., Schreier, P., HRGC & CC, in press.

12. Idstein, H., Herres, W., Schreier, P., J. Agr. Food Chem., in press.

13. Schreier, P., Idstein, H., Herres, W., in Analysis of Volatiles (Schreier, P., ed.), 293, W. de Gruyter, Berlin, 1984.

14. Siemens Sichromat 2 with "life" column switching technique, Siemens, Karlsruhe, FRG (cf. Oreans et al., in Analysis of Volatiles (Schreier, P., ed.), 171, W. de Gruyter, Berlin, 1984.

15. Herres, W., Neumann, P., to be published.

APPLICATION OF DIRECT CARBON-13 NMR SPECTROSCOPY IN THE ANALYSIS OF VOLATILES

Karl-Heinz Kubeczka
Lehrstuhl für Pharmazeutische Biologie der Universität
Würzburg, Mittlerer Dallenbergweg 64, D-8700 Würzburg

Viktor Formácek
Bruker Analytische Meßtechnik GmbH, 7512 Rheinstetten-Forchheim

The analysis of complex multicomponent mixtures like natural
flavours or essential oils requires the application of modern
physico-chemical methods and sophisticated instrumentation.
Nowadays, different chromatographic techniques are predominant
in the analysis of volatiles; in particular, gas chromatographic methods wich result in the separation of the mixture into
individual components play an important role. Furthermore, the
techniques allow to determine, qualitatively and quantitatively, the individual fractions obtained in the chromatogram. However, if changes in the composition occur in the course of chromatography due to thermal or other influences, the results are
considerably in error and may be useless.

As an alternative to chromatographic evaluation of volatile
compounds, methods that provide direct information about the
composition of a particular sample without previous separation
of components may be utilized, i.e. spectroscopic methods. For
example, infrared spectroscopy has been applied (1,2) to obtain
information about the composition of essential oils. Disadvantages, however, have to be mentioned; first, sensitivity and
selectivity of infrared spectroscopy are low in the case of
multicomponent mixtures, and secondly, extremely difficult

problems are confronted, when attempting to quantitatively
measure individual component concentrations. If a mixture has
a make up, in which the main component is present in a high
proportion, then a separation of minor constituents from the
major component is practically impossible. The accompanying
substances can be seen only with difficulty and as far determination of structure is concerned, cannot be seen at all.

Theoretical considerations concerning the applicability of
other spectroscopic methods in the analysis of flavours and
essential oils have induced us to turn to nuclear magnetic resonance spectroscopy (3,4,5,6). The ^{1}H-NMR spectrum of a complex mixture results from the spin to spin coupling of the individual nuclei over complex and relatively broad bands, which
often overlap because of minimal differences in shift. The entire available resonance region of protons includes approximately 0 - 15 ppm. For this reason, the method cannot be used for
complex mixtures like essential oils.

The case is different for ^{13}C-NMR spectroscopy. In the ^{13}C-NMR
spectrum of a mixture, the singlet signals of the ^{13}C nuclei
are usually well resolved and despite relatively low sensitivity, the analytical possibilities of ^{13}C-NMR turn out to be
much better than those of the ^{1}H-NMR , especially if sufficiently concentrated solutions of the analyzed mixtures are available. The natural occurrence of the ^{13}C nucleus is only 1.1%.
Thus the probability is extremely small, that two ^{13}C atoms
will be found next to one another in a molecule. A homo-nuclear
spin-spin coupling in all practibility does not then occur. The
presence in the spectrum of the important ^{13}C-^{1}H couplings,
which are needed to clarify exact structure, can be eliminated
with the use of a disturbance field, thus simplifying the
spectrum. At the same time the signal to noise ratio and the
sensitivity of the measurement are increased.

From this so called proton broad band decoupling one obtains
a relatively simple spectrum. An overlapping of individual

resonance signals occurs rather seldom due to a considerably larger chemical shift and minimal signal width. The region in which one can work successfully with ^{13}C-NMR spectroscopy is from 0 - 240 ppm.

The main constituent of a mixture, which is present in a high proportion like eugenol in the essential clove oil with about 85% can be recognized after a single puls of 1 second, using a 90% solution in benzene-d$_6$. After several minutes accumulation, the signal to noise ratio will be enhanced significantly (Fig.1). For the determination of further oil components present in yet lower concentrations, a longer accumulation time is needed. The qualitative analysis of an individual component is based upon the comparison of the spectrum of the mixture with spectra of pure reference compounds (Fig.2). These spectra should be recorded, if possible, under identical conditions of solvent, temperature, lock substances etc. This assures that differences in the chemical shift for the individual ^{13}C-NMR lines of the reference substances and of the individual components of the mixture are negligible. It seems, therefore, absolutely necessary to establish one's own data bank of pure compounds and to use published spectra only for references.

To investigate the influence of other constituents of the mixture on the chemical shifts, the spectra of about 50 essential oils have been evaluated. The deviations of the individual NMR shifts of a component were less than 0.5 ppm except of the carbonyl resonance lines.

Benzene-d$_6$ was always used as the solvent and the lock substance due to good solubility of all species investigated in this solvent. Moreover, benzene is chemically inert, which is especially important when investigating sesquiterpenes that sometimes easily undergo proton catalyzed rearrangements. Further, the C_6D_6 ^{13}C-NMR lines appear at a convenient position in the spectrum. The central line of the C_6D_6 triplet has been used as a chemical shift reference at 128.0 ppm relative to TMS.

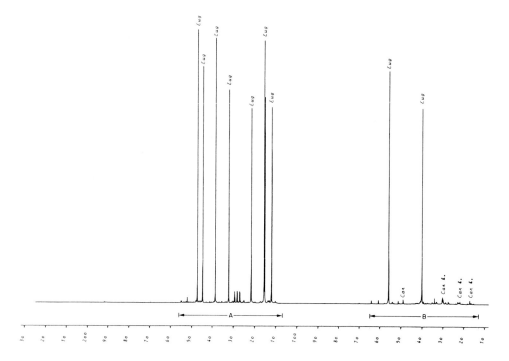

Fig. 1   Proton broad-band decoupled ^{13}C-NMR spectrum of the
           essential clove oil with about 85% of eugenol

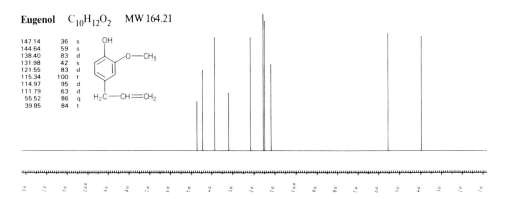

Fig. 2   Proton broad-band decoupled ^{13}C-NMR spectrum of pure
           eugenol

As an example of a more complex mixture, we may consider coriander oil with linalool as a main constituent and in addition
with small amounts of different monoterpene hydrocarbons,
esters, alcohols and ketones. Similarily, here the main component linalool is recognized quickly (Fig.3) and can be easily
identified through use of a reference spectrum. After a 12
hour accumulation time, components present at less than 1%
concentration continue to be observed and may be identified.
For recognition of these, a horizontal as well as a vertical
expansion of the spectrum is advantageous. The region from
150 - 105 ppm (A), 80 - 36 ppm (B) and 36 - 5 ppm (C) should
be plotted separately (Fig.4, 5 and 6) in order to bring out
the resonance characteristics of aliphatic carbon atoms as
well as aromatic C-atoms and double bonds.

After comparing the expanded parts of the spectrum and the
data print out to data obtained with references, the qualitative results may be tabulated.

In the case of very complex mixtures, a preseparation of the
sample into several less complex fractions - for example by
preparative scale HPLC - seems to be advantageous to enhance
selectivity and sensitivity of the method. By this procedure
we have been able to analyze even components present in trace
amounts with less than 0.1% in the original mixture without
any difficulty.

Along with the possibility of qualitative analysis by ^{13}C-NMR
spectroscopy, we have also attempted to use this technique for
quantitative analysis. For this purpose it is necessary to
study the broad band proton decoupled spectra in detail. These
relatively simple spectra are characterized by small, almost
equally broad signals, which in the simplest cases may be
assigned to individual C-atoms. In the following, we restrict
our comments to results obtained from experiments with such
broad band decoupled ^{13}C-NMR spectra (7).

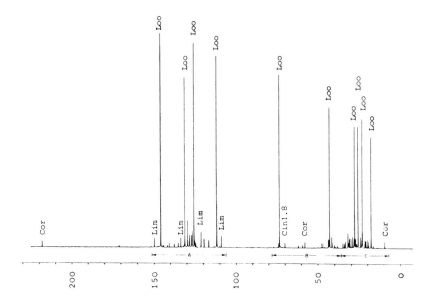

Fig. 3 Proton broad-band decoupled ^{13}C-NMR spectrum of
coriander oil with linalool (Loo) as a main constituent

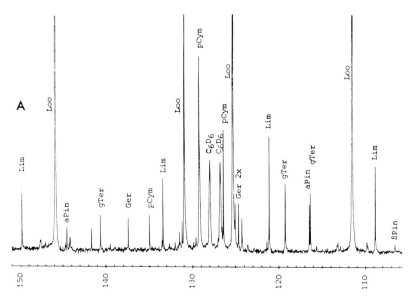

Fig. 4 ^{13}C-NMR spectrum of coriander oil; expanded aromatic
region

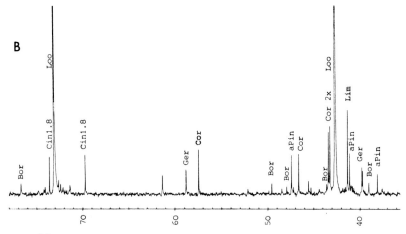

Fig. 5 ^{13}C-NMR spectrum of coriander oil; expanded aliphatic
          region 80 - 36 ppm

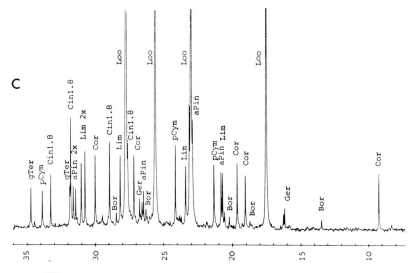

Fig. 6 ^{13}C-NMR spectrum of coriander oil; expanded aliphatic
          region 36 - 5 ppm

Bor	= Borneol		Lim	= Limonene
Cin1.8	= 1,8-Cineole		Loo	= Linalool
Cor	= Camphor		aPin	= alpha-Pinene
pCym	= para-Cymene		ßPin	= beta-Pinene
Ger	= Geraniol		gTer	= gamma-Terpinene

In contrast to ^{1}H-NMR spectroscopy, in these experiments integration of individual signals has not proven useful for quantitative determinations. Much smaller errors are found when measuring signal height. Due to the large vertical dynamics of ^{13}C-NMR spectra of natural mixtures, the signal to noise ratio of components present in low concentrations is extremely disadvantageous. The humps of week signals disappear in the noise and the integral values are to low. The noise effects are less important, if the height of the signal is evaluated.

The number of factors which influence the quantitative interpretation of ^{13}C spectral data is much larger than in the case of ^{1}H-NMR spectra. Basically, these factors may be grouped into two classes:

1. effects which result due to the properties of analysis substances and
2. effects which arise due to the spectrometer used.

In the first case, the effects can be understood as resulting from differences in relaxation times and the nuclear Overhauser effect (NOE) of individual nuclei. Signal intensities of ^{13}C atoms, which are not directly bound to protons, as for example quarternary C-atoms and C-atoms of carbonyl groups, are greatly effected due to both of these phenomena. Such nuclei are characterized by a relatively long relaxation time and a low NOE. As a result, signal intensity is reduced.

An opposite effect of relaxation time and of the NOE in the case of methyl groups leads to an approximate compensation in both effects.

Special techniques which suppress the effects of relaxation time and NOE (8-14) are not applicable when attempting to quantitatively analyze mixtures of many components, either because measurement times are extended unacceptably, or because they produce undesirable broadening of lines. Thus, errors arising in quantification due to the substances involved, may

only be eliminated by ignoring the respective resonance signals.

In the process of digitazing the resonances, variations in the
intensity of a individual signal may arise due the position
of the data points (Fig.7). Such variations of the actual
value may be reduced by mathematical treatment of the interferogram before Fourier transformation as for example by ex-
ponential multiplication. Such a mathematical treatment of
measured values is strongly recommended for all quantitative
analyzes and we have in all cases followed such a procedure.

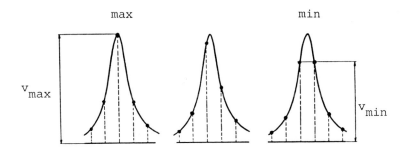

Fig. 7   Variation of the actual value v due to the position
          of the data points during the process of digitazing

The average signal intensity $\bar{I}$ of individual resonance lines
- measured as signal height - may be used as a basis for the
quantitative determination of a substance by ^{13}C-NMR -spectro-
scopy. One must realize, that each measured signal results
from a single C-atom. In calculation of $\bar{I}$, it is necessary to
consider that in the case of a symmetrical molecule, several
resonances may overlap. Moreover, due to the differing carbon
content of individual components one must realize that only
the total carbon of a molecule may be determined with 13C NMR
spectroscopy. Therefore, the product of average signal intensity and molecular weight must be taken. To determine the percent content of a component n of a mixture, which consists of
x components, the following equation holds:

$$\%n = \frac{MW_n \cdot \bar{I}_n \cdot 100}{MW_a \cdot \bar{I}_a + MW_b \cdot \bar{I}_b \ldots + MW_n \cdot \bar{I}_n \ldots + MW_x \cdot \bar{I}_x}$$

$\%n$ = percentage of a component in the mixture
$\underline{MW}$ = molecular weight
$\bar{I}$ = average signal intensity of a component, measured as signal height

Since for such a calculation all components must be known - which relatively seldom occurs - often the method of internal standardization must be applied as an alternative to quantitative determination of individual compounds. For this purpose a pure aliquot of a compound, which is already present in the mixture is added as a standard to the sample to be analyzed. The increase in signal intensities allows accurate estimation of the percent content for the standard component and for other known mixture components. Qualitative results obtained with both methods for an analysis of fennel oil are presented in table 1. In addition, we have included for comparison re-

Table 1   Comparison of quantitative analyses of fennel oil by 13C NMR spectroscopy and gas chromatography

Compound	$\%^1$	$\%^2$	GC $\%$
trans-Anethole	25.6	25.2	21.36
Limonene	25.4	25.0	26.80
Estragol	2.5	2.5	2.07
Fenchone	17.0	16.7	14.61
p-Cymene	1.2	1.2	1.73
α-Pinene	9.3	9.2	10.45
ß-Pinene	2.8	2.8	3.22
Camphene	0.4	0.4	0.49
1.8-Cineole	2.1	2.1	2.47
Myrcene	2.5	2.5	2.69
α-Phellandrene	10.1	9.9	9.72
ß-Phellandrene	1.1	1.1	1.09
Total	100.0	98.6	96.70

$\%^1$ = NMR determination by the 100% method; $\%^2$ = NMR determination by internal standardization

sults, obtained for the same mixture with capillary gas chromatography. One observes that the 100% determination procedure and the internal standard procedure result in comparable determinations which also agree with data obtained by GLC.

With recent development of an appropriate computer program, we have now been able to automate the evaluation of ^{13}C-NMR spectra of mixtures in both qualitative and quantitative studies. Details of the computerization will be reported in the near future.

Conclusion

As the experiments have shown, it is possible to determine both qualitatively and quantitatively the components of complex mixtures such as essential oils with the help of ^{13}C-NMR spectroscopy. The analysis by means of ^{13}C-NMR spectroscopy may be especially effective in the determination of the composition of natural flavours or essential oils whose components are partially thermolabile or non-volatile and therefore cannot be directly analyzed by GLC. A further advantage of this method is the possibility of automated data analysis.

References

1.   Bellanato, J., Hidalgo, A.: Infrared analysis of essential oils, Heyden & Son, London 1971
2.   Farnow, H.: Dragoco Report 15, 27 (1968)
3.   Formáček, V.: Thesis, Wuerzburg 1979
4.   Formáček, V., Kubeczka, K.-H.: Application of 13C NMR spectroscopy in the analysis of essential oils; in Kubeczka, K.-H. (ed.): Vorkommen und Analytik ätherischer Öle, Thieme Verlag, Stuttgart 1979
5.   Formáček, V., Kubeczka, K.-H.: 13C NMR analysis of essential oils; in Margaris, N., Koedam, A., Vokou, D. (eds.): Aromatic

Plants: Basic and Applied Aspects, Martinus Nijhoff Publishers, The Hague. Boston. London 1982

6. Formáček, V., Kubeczka, K.-H.: Essential oils analysis by capillary gas chromatography and carbon-13 NMR spectroscopy, John Wiley & Sons, Chichester. New York. Brisbane. Toronto. Singapore 1982

7. Formáček, V., Kubeczka, K.-H.: Quantitative analysis of essential oils by 13C NMR spectroscopy; in Kubeczka, K.-H. (ed.): Ätherische Öle - Analytik, Physiologie, Zusammensetzung, Thieme Verlag, Stuttgart. New York 1982

8. Feeney, J., Shaw, D., Pauwels, P.S.J.: Chem. Commun. 554 (1970)

9. Gansow, O.A., Schittenhelm, W.: J.Am.Chem.Soc. 93, 4294 (1971)

10. Freeman, R., Hill, H.D.W., Kaptein, R.: J.Mag. Res. 7, 327 (1972)

11. Freeman, R., Hill, H.D.W.: J.Mag. Res. 5, 278 (1971).

12. Levy, G.C., White, D.M., Anet, F.A.L.: J.Mag. Res. 6, 453 (1972)

13. Lamar, G.N.: J.Am.Chem.Soc. 93, 1040 (1971).

14. Barcza, S., Engstrom, N., Arner, S.: Chem.Soc. 94, 1762 (1972)

APPLICATIONS

APPLICATION OF HIGH RESOLUTION CAPILLARY COLUMNS ON FLAVOR AND
FRAGRANCE ANALYSIS

Takayuki Shibamoto
Department of Environmental Toxicology
University of California, Davis, CA 95616

Introduction

History of fragrances

The fragrances of certain plants have attracted people for
many years. The Egyptians had extensive knowledge of aromatic
plants, as they illustrated by reference to myrrh, cinnamon,
galbanum, and many similar materials in the Papyrus Ebers
dating from 2000 B.C. Fragrance was used not only for
perfume, which was said to enhance the beauty and charm of
Cleopatra, but also in ceremonies, such as funerals. The
perfumes developed by Egyptians were later adopted and used by
both the Greeks and the Romans. Many ingredients which are
used today already appeared in their recipes for fragrances:
Pistacia lentiscus, cinnamon, peppermint, and Convolvulus
scoparius. The detection of natural animal fat in their
perfume indicates this was the substance used to extract
essence from the plants.

The use of fragrance materials by the people of India goes
back to the earliest period of their history. India is rich
in trees with perfumed woods such as sandalwood and patchouli.
Their essential oils were widely used for embalming the bodies
of the princesses of Ceylon in the 9th century. The same wood
oils were also used as food preservatives.

The Arabic world did much to develop chemistry between the 9th
and 12th centuries. They discovered the process of distillation which allowed the preparation of the volatile oils of
flowers. The flowers most prized by the Arabs for their
perfumes were jasmine, lily, orange flower, sweet basil, wild
thyme, lofas, crocus, saffron, narcissus, and the flowers of
various kinds of beans. Oils from animals such as musk and
civet were also first used in Arabic perfumes.

In Italy, perfumery began in the early 16th century, when
Venice became a major trade center. At this time, the royal
family and the nobility began to use perfumes extensively in
their palaces. In 1508, the first laboratory for perfumes was
founded in Florence. During the 17th and 18th centuries, many
laboratories produced numerous scents, which were placed in
tiny bottles in small boxes or cases in France, and distributed throughout the world. From that time on, France has
been known as the world center of perfumes.

Today, there are about 200 well-known perfumes sold in the
world. Until the end of the 19th century, most perfumes
consisted only of natural products. The development of
organic chemistry by French and German scientists after the
war of 1870, created the synthetic odorant industry. Advances
in synthetic chemistry allowed the creation of tremendous
numbers of new fragrances.

Analysis of fragrance

Analysis of the major natural oils remains the first step in
fragrance studies. Many scientists have tried to analyze
plant essences in order to pinpoint the chemicals which give
the characteristic fragrance to each plant. The vanillin
found in vanilla beans and cinnamic aldehyde from cinnamon

were synthesized in 1874 and 1884, respectively. Since the
beginning of the 20th century, the major constituents of plant
essences have been isolated and identified by a conventional
wet chemical method. Guenther (1952) summarized the nature
and constituents of various essential oils in his six volume
reference work.

Since James and Martin succeeded in making a prototype gas
chromatograph in 1952, this instrument has become the most
powerful tool in the analysis of flavors and fragrances. In
the 1960s, the development of high resolution columns and gas
chromatography/mass spectrometry techniques permitted the
identification of a tremendous number of the flavor and fragrance constituents of various substances such as essential
oils, cooked foods, fruit essences, and flower extracts.
Column development in particular seems to be an area where
continued research has advanced flavor and fragrance
analysis. More novel constituents of flavor and fragrance
materials may be isolated and identified following the
development of high resolution gas chromatographic columns.

Gas Chromatographic Columns

The most important and crucial part of a gas chromatograph is
the column, since this is where separation occurs. Early
studies on flavor and fragrance analysis were done using
packed columns. Many liquid phases were used, but the resolution obtained by a packed column is quite low compared with
the more recent capillary column. For example, the gas chromatogram of peppermint oil (Figure 1) obtained by a 2 m × 3 mm
i.d. stainless steel column packed with 5% Carbowax 20M on
Chromosorb W (80/100) shows only several peaks. The same oil
was found to contain more than one hundred components when a
high resolution glass capillary column was used in later
analyses (Figure 2).

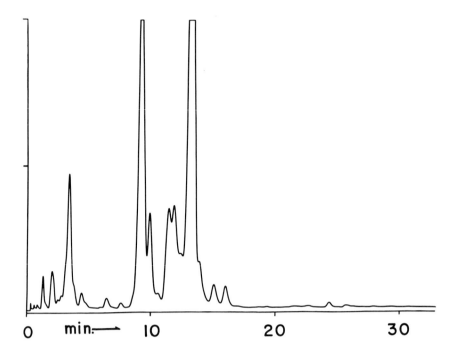

Figure 1   Gas chromatogram of peppermint oil obtained by a
stainless steel column packed with 5% Carbowax 20M on
Chromosorb W (80/100).

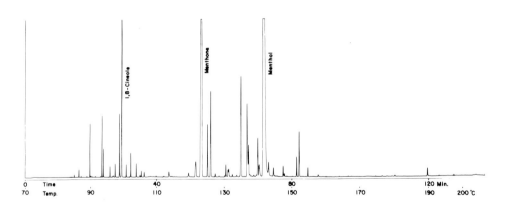

Figure 2   Gas chromatogram of peppermint oil: 50 m × 0.28 mm
i.d. glass capillary column coated with Carbowax 20M and
programmed from 70° to 200°C at 2°/min; nitrogen carrier gas
flow rate 15 cm/sec.

In Figure 1, the three major peaks were monoterpene hydrocarbons, menthone, and menthol. In comparison, Figure 3 shows a gas chromatogram of peppermint oil obtained by a classic stainless steel <u>capillary</u> column coated with Carbowax 20M. In 1960, this class of column was recognized as the highest resolution column. α-Pinene and β-pinene were separated first by this column. Figure 2 indicates that a glass capillary column can separate d-limonene and 1,8-cineol; further, base line separation can be seen in the three peaks at menthone. Many single peaks in Figure 3 split into multiple peaks in Figure 2. Figure 4 shows a gas chromatogram of the same peppermint oil as Figures 1, 2, and 3, obtained with a more recent fused silica capillary column coated with Carbowax 20M. The most apparent difference between glass and fused silica columns is that fused silica gives a tail-free peak for menthol.

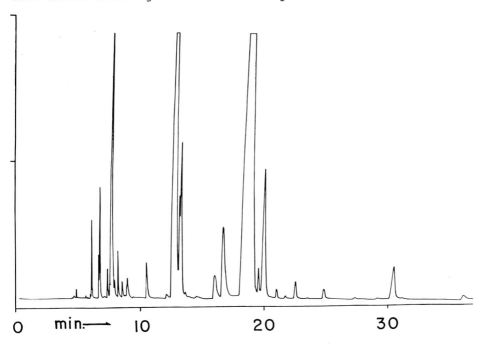

Figure 3   Gas chromatogram of peppermint oil: 45 m × 0.5 mm i.d. stainless steel capillary column coated with Carbowax 20M and programmed from 100°C to 175°C at 3°/min; nitrogen carrier gas flow rate 15 cm/sec.

238

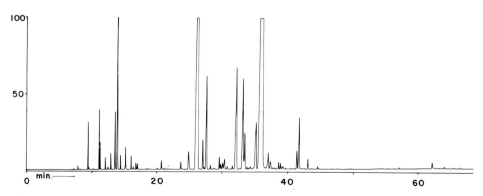

Figure 4 Gas chromatogram of peppermint oil: 50 m × 0.23 mm
i.d. fused silica capillary column coated with Carbowax 20M
and programmed from 70°C to 200°C at 2°/min; nitrogen carrier
gas flow rate 15 cm/sec.

As noted above, the invention of a high resolution capillary
column permitted significant progress in flavor and fragrance
analysis. Pursuing the development of a yet higher resolution
column will shed light on the remaining unknown constituents
of many materials.

Preparation of a High Resolution Capillary Column

Column coatings

A revolutionary column called the open tubular column was
invented in 1958 (1); following this invention, many researchers worked at developing coating techniques to obtain high
resolution capillary columns. There are two general methods
for column coating, and a summary of these techniques may be
found in Jennings (2). Resolution was significantly increased
with the invention of the glass capillary column. The method
of drawing glass capillary tubing was invented in 1960, however, there were many problems involved in preparing a stable
liquid phase coating on the glass surface in the early years

of its use. As a result of intensive work on coating techniques by many researchers -- e.g., work on the use of wetting
agents (3) -- today stable high resolution glass capillary
columns are available commercially.

The dynamic coating method is most commonly used. A solution
containing a liquid phase (5-10%) in a suitable low boiling
solvent is forced through the column by pressurized nitrogen
gas (4). Later, even more satisfactory results were obtained
by utilizing more concentrated (i.e., more viscous) coating
solutions (5). In this method, mercury is used to sweat the
solution out from a column. One advantage of the dynamic
coating method is that the apparatus is simple and relatively
inexpensive. It is, however, difficult to obtain the desired
coating thickness by this method. Static coating techniques
developed largely by Verzele and his associates (6, 7) allow
us to obtain desired coating thickness more easily. In this
method, the column is filled with a dilute solution (0.3-0.5%)
of liquid phase in a suitable low-boiling solvent, and one end
is sealed. The column is slowly introduced into a heater from
the other end and the solvent is evaporated off, leaving a
thin film of liquid phase.

The influences of coating thickness on gas chromatographic
retention times seem to be important in determining standard
retention indices for chemicals. We prepared five columns
with different coating thicknesses (prepared by static method)
in order to investigate the influence of coating thickness on
Kovats Index I (8). The characteristics of each column are
shown in Table I. The I differences (ΔI) for alcohols, aldehydes, and ethyl esters measured on columns 2 and 5 (Table I)
at 80°C and 120°C are shown in Table II. The results indicate
that ΔI is larger at 80°C than at 120°C for all chemicals
tested. It was concluded that a column of large d_f (coating
thickness) at high temperature should be used to minimize ΔI.

Table I   Characteristics of Columns Used for the Experiment of Kovats
Index.

Column No.	Conc. of liquid phase (mg/ml)	Coating thickness (d_f)		K_{120}^c	No. of theoretical plates	
		Measured[a] (μm)	Theoretical[b] (μm)		at 80°C	at 120°C
1	1	0.041	0.058	0.51	130,000	230,000
2	2	0.089	0.116	0.90	116,000	176,000
3	4	0.200	0.232	1.66	107,000	145,000
4	8	0.320	0.464	2.78	103,000	125,000
5	12	0.450	0.696	4.11	100,000	116,000

[a] Actually measured by scanning electron microscope.

[b] Calculated using the equation $d_f = 5cr/p$ (d_f = thickness of the liquid phase film in μm; r = inner radius of the tube in cm; p = density of the liquid phase; c = concentration of liquid phase, mg/ml, in the coating solution).

[c] The partition ratio measured at 120°C for C_{14} n-paraffin hydrocarbon.

Table II   The I Differences (ΔI) for the Alcohols, Aldehydes, and Ethyl
Esters Measured by Column 2 and Column 5 at 80° and 120°C.

Compound	ΔI at 80°C	ΔI at 120°C
Alcohols		
n-pentanol (C-5)	50	34
n-hexanol (C-6)	54	34
n-heptanol (C-7)	57	38
n-octanol (C-8)	65	42
Aldehydes		
n-heptanal (C-7)	42	26
n-octanal (C-8)	46	30
n-nonanal (C-9)	53	36
n-decanal (C-10)	59	40
Ethyl esters		
ethyl-n-hexanoate (C-6)	37	33
ethyl-n-heptanoate (C-7)	46	37
ethyl-n-octanoate (C-8)	52	43
ethyl-n-nonanoate (C-9)	57	46

Fused silica capillary columns

Recently, a highly flexible, fused-silica capillary column has
been developed and is commercially available. Because of its
flexibility, installation of a column does not require special
skill. The most striking feature of the fused silica capillary column is the inertness of the column surface. For
example, Figure 5 shows the difference in the peak shapes of
free fatty acids between a glass column and a fused silica
column. Fused silica columns show tail-free peaks with free
fatty acids, alcohols, phenols, and amines. Lipsky and his
associates compared two types of columns, natural quartz and
fused silica, and obtained superior results most often from
fused silica columns (9). Their chromatograms showed that
fatty acids do not tail on a fused silica column but do on a
natural quartz column, when both are coated with Carbowax 20M.
Another example is shown in the gas chromatogram of California
hop oil taken by a fused silica capillary column (Figure 6).

It must be noted, however, that in our series of routine
flavor and fragrance analyses, we have found some contradic-
tory problems in the use of a fused silica capillary column.
The acid peaks are easily recognized on a chromatogram from a
glass capillary column because of tailing. This fact makes
GC/MS analysis easier. Some acids give identical MS fragmen-
tations to their ester derivatives which do not give tailing
peaks on a chromatogram obtained by a glass capillary column.
The acids and their ester derivatives can, therefore, be
distinguished using the tailing phenomenon on a chromatogram
obtained by a glass capillary column.

Manufacturing a fused silica capillary column is much more
difficult than manufacturing a glass capillary column;
generally, the former are too difficult to manufacture in
private laboratories due to the requirement of high

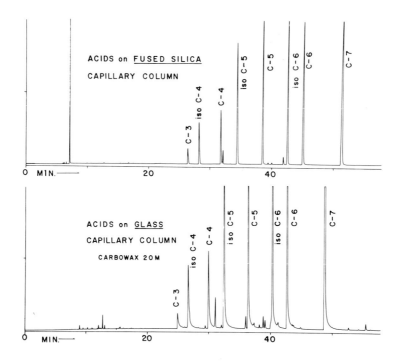

Figure 5   Gas chromatograms of a n–fatty acids mixture on a
fused silica capillary column (top, 50 m × 0.23 mm i.d.) and
on a glass capillary column (bottom, 50 m × 0.28 mm i.d.),
both coated with Carbowax 20M.

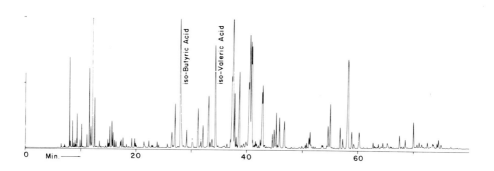

Figure 6   Gas chromatogram of California hop oil obtained by
a fused silica capillary column coated with Carbowax 20M
(50 m × 0.23 mm i.d.)

temperature (1800-2000°C) to draw the column. The outside
surface of the column must be coated with polyimide to
increase flexibility.

Figure 7 shows a diagram of the drawing process and the
outside coating. Double coating on the outside surface
increases the column's flexibility. Once a stable,

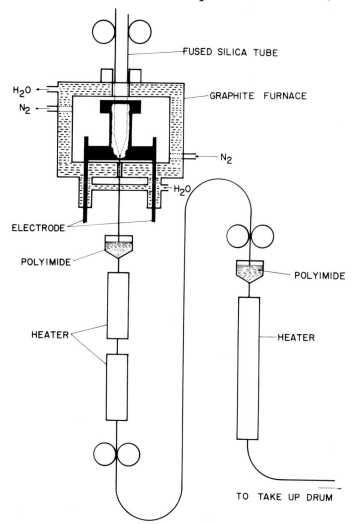

Figure 7   Diagram of the drawing and the outside coating
instrument.

fused-silica capillary column is obtained, inside coating with a liquid phase by the static method is fairly easy. Figure 8 shows the apparatus prepared for liquid phase coating for fused silica columns. The heating zone can be straight, which makes the system considerably simpler than the one for a glass capillary column.

Several manufacturers are now supplying various fused silica capillary columns. There are generally two kinds of capillary columns. One is a wall-coated column in which a liquid phase is simply coated on the column surfaces. The other is one in which a liquid phase is chemically bonded to the column surfaces. The preparation of the bonded phase column is not yet widely known and has not been reported in the literature in any detail. According to our experience, a bonded phase column stands up under higher temperatures than does a coated

Figure 8   Photograph of the liquid phase coating instrument

phase column. However, a coated phase column is superior to a
bonded phase column in the reproducibility of retention time.
Resolution is equal for the two columns.

Examples of Applications

Gas chromatography is used for the analysis of almost any type
of organic chemical. The analysis of natural plant components,
air pollutants, pesticide residues, flavor and fragrance chemicals, and biological substances such as plasma cholesterols
are now performed routinely using a gas chromatograph. Today,
the analysis of volatile chemicals is highly dependent upon
gas chromatography. Tremendous numbers of volatile chemicals
have been identified since the invention of gas chromatography. Introduction of a high resolution glass capillary column
opened a new world to the analytical chemists. Gas chromatography demonstrated that essential oils consist of hundreds of
chemicals. For example, the oil of Schinus molle L. contains
more than two hundred chemicals (Figure 9). Forty-six compounds were identified in the above oil using a 50 m × 0.28 mm
WCOT glass capillary column.

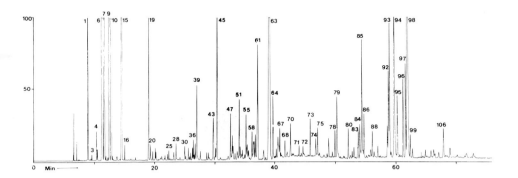

Figure 9   Gas chromatogram of oil of Schinus molle L.
obtained by a fused silica capillary column (50 m × 0.23 mm
i.d.) coated with Carbowax 20M.

246

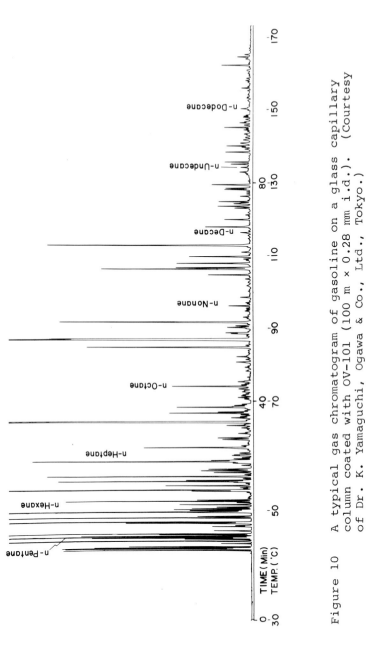

Figure 10   A typical gas chromatogram of gasoline on a glass capillary column coated with OV-101 (100 m × 0.28 mm i.d.).   (Courtesy of Dr. K. Yamaguchi, Ogawa & Co., Ltd., Tokyo.)

We now know that gasoline contains more than 300 components
(Figure 10).

Analysis of food flavors has been conducted most actively
using gas chromatography with capillary columns. Gas chromatography is an ideal technique for analyzing flavor volatiles
of foods and beverages, because the basic principle of gas
chromatography is the separation of mixtures in the vapor
phase. When we enjoy flavors of foods or beverages, we are
enjoying the vapor phase of those flavor compounds. Figure 11
shows a typical gas chromatogram of coffee volatiles.

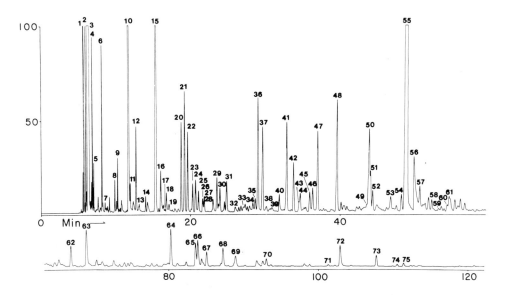

Figure 11   Gas chromatogram of volatiles obtained from
Columbian coffee on a fused silica capillary column (50 m ×
0.23 mm i.d.) coated with Carbowax 20M. Oven temperature was
programmed from 70°C to 200°C at 2°/min; helium carrier gas
flow rate 23 cm/min (courtesy of Dr. K. Yamaguchi, Ogawa &
Co., Ltd., Tokyo).

Numerous sulfur-containing compounds were identified in coffee volatiles using glass capillary columns. It has been suggested that most trace amounts of sulfur-containing compounds do not go through stainless steel columns due to the active surfaces, however thiazoles can now be identified in meat volatiles using a glass capillary column. Figure 12 shows a typical gas chromatogram of meat broth volatiles.

It is well-known that the heat treatment of foods produces numerous volatiles. For example, raw milk possesses virtually no flavor but heated milk contains a tremendous number of volatile chemicals as is shown in Figure 12.

The invention of gas chromatography also produced a revolution in perfumery research. Until the development and dissemination of gas chromatography techniques, the analysis of perfumery materials depended totally upon a trained perfumer's nose. Even a highly skilled perfumer, however, can identify no more than 20-30 components of a perfume. A high resolution capillary column shows that perfume may contain hundreds of chemicals. Figure 13 shows a gas chromatogram of a commercial cologne obtained by a fused silica capillary column. Since there is no way to express odors in terms of physical measurements, a gas chromatographic pattern is one of the most convenient and useful ways to depict those odorous mixtures. A high resolution capillary column is widely used in the perfumery industries today because it offers an ideal pattern for the recording of the nature of perfumery products.

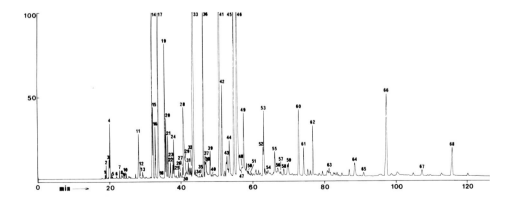

Figure 12   A typical gas chromatogram of beef broth volatiles
on a glass capillary column (50 m × 0.28 mm i.d.) coated with
Carbowax 20M.

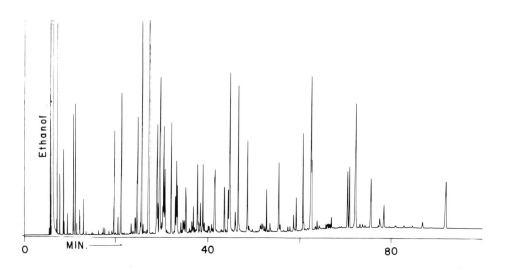

Figure 13   Gas chromatogram of a commercial cologne on a
fused silica capillary column (50 m × 0.23 mm i.d.) coated
with Carbowax 20M (courtesy of Dr. K. Yamaguchi, Ogawa & Co.,
Ltd., Tokyo).

References

1. Golay, M.J.E.: Theory and practice of gas liquid partition chromatograhy with coated capillaries, Gas Chromatography, Coates, V.J., Nobels, H.J. Fagerson, I.S. (eds.), Academic Press, New York 1958, pp 1-13.

2. Jennings, W.G.: Gas Chromatography with Glass Capillary Columns, Academic Press, New York 1982, pp 39-51.

3. Metcalfe, L.D., Martin, R.J.: Anal. Chem. $\underline{39}$, 1204 (1967).

4. Dijkstra, G., De Goey, J.: Gas Chromatography, Desty, D.H. (ed.), Butterworths, London 1958, pp 56-58.

5. Schomburg, G., Husmann, H.: Chromatographia $\underline{8}$, 517 (1975).

6. Bouche, J., Verzele, M.: J. Gas Chromatogr. $\underline{6}$, 501-505 (1968).

7. Verzele, M., Verstappe, M., Sandra, P., van Luchene, E., Vuye, A.: J. Chromatogr. Sci. $\underline{10}$, 668-673 (1972).

8. Shibamoto, T., Harada, K., Yamaguchi, K., Aitoku, A.: J. Chromatogr. $\underline{194}$, 277-284 (1980).

9. Lipsky, S.R., McMurray, W.J., Hernadez, M., Purcell, J.E., Billeb, K.A.: J. Chromatogr. Sci. $\underline{18}$, 1-9 (1980).

Acknowledgment

The author thanks Drs. K. Yamaguchi and S. Mihara (Ogawa & Co., Ltd., Tokyo) for their useful advice in the preparation of this manuscript.

THE ANALYSIS OF ODOR-ACTIVE VOLATILES IN GAS CHROMATOGRAPHIC EFFLUENTS

Terry E. Acree and John Barnard

Cornell University
NYS Agr. Exp. Sta.
Geneva, New York 14456, USA

David G. Cunningham

Food and Drug Administration
200 C Street SW
Washington, DC 20204 USA

Introduction

The "sniffing" of gas chromatographic effluents in order to associate odor-activity with a particular eluate is an inevitable part of the study of volatile flavor. It is not unusual to observe the elution of an odor from a gas chromatographic column at a retention time when no detectable chemicals can be seen using more objective measurements such as flame ionization (1). This is because the human nose is a very sensitive detector. In addition, the nose is a very selective detector responding to only a few of the chemicals in most natural products. This selective sensitivity combined with the physiological problems of adaptation and fatigue create special difficulties for the analyst.

Analysis of Volatiles
© 1984 Walter de Gruyter & Co., Berlin · New York – Printed in Germany

The exact day when man began to sniff gas chromatographic effluents is not known but it probably occurred soon after the invention of gas chromatography in 1952 (2). The predominant problems in the sensory analysis of gas chromatographic effluents were outlined in 1964 (3) by Fuller et al., and Figure 1 presents a graphic display of the central concern: gas flow.

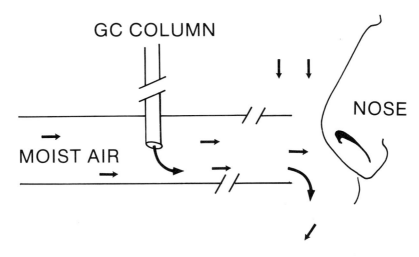

Figure 1   Diagram showing the three different gas flows required to bring gas chromatographic effluents to human subjects for sensory analysis.

The gas eluting from the chromatograph has a high linear velocity but an extremely low volume flow or mass transfer. This is especially true when narrow bore capillary columns are used. The problem is solved by the dilution of the eluent gas with a much larger volume-flow of air, preferably air humidified and cooled to a level appropriate to the delicate membranes in the human nose (4,5,6). Finally, all of the gases in the neighborhood of the nose must be removed or diluted with an effective exhaust and ventilation system. Again this is particularly true when evaluating the narrow bands eluting from high resolution capillary columns.

Once these mechanical problems have been solved, a suitable scenario for the recording of the sensory perceptions experienced by the human detector must be devised. If nominal scaling (descriptive analysis), or the recording of the odor quality or character eluting from a gas chromatograph is the type of information being sought, these results can be collected on a moving paper recorder, an audio recorder, or a real-time computer (7-13). The nominal scaling of gas chromatographic effluents is not unique and a proper description of the technique can be found elsewhere (14), but a concise statement of the requirements for nominal scaling is that a person: 1, has encountered the odor before; 2, has made a connection between an odor and a name; and 3, has an aid in recalling that name (15).

However, the problem of associating an intensity with an odor eluting from a gas chromatograph does have some special restrictions. The dynamics of gas chromatography preclude the use of discriminant sensory analysis or any comparative procedure. Thus, the common approach has been to use ordinal scaling with a small number, 3 to 9, of divisions (11,12,16) with an occasional use of magnitude estimation (9). A visually impressive result was provided by the application of a principle called cross modal matching, in which the intensity of one sensory perception, e.g. odor, is associated with another perceived sensation, e.g. the position of a lever (17). The lever position was measured electronically, digitized and the results used to generate an odor intensity gas chromatogram.

Sensory Saturation

Any of the techniques referenced above, when judiciously applied, have produced useful information about the odor activity of gas chromatographic effluents. However, the occurrence of sensory saturation and adaptation is more common during the sniffing of chromatographic effluents than during other types of sensory experiments. This is because these effluent chemicals are 100% volatilized. A solution with a faint odor when injected into a gas chromatograph can produce an overpowering odor at the other end of the column. A typical example of this was a report that beta damascenone had no odor when it was sniffed eluting from a gas chromatographic column (18); clearly an example of sensory saturation since one picogram of this compound can be easily detected eluting from a gas chromatographic column.

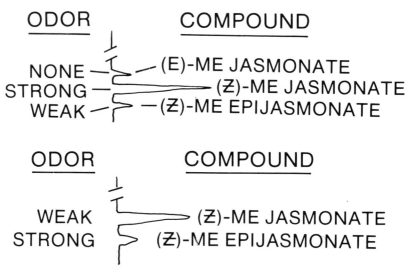

Figure 2   A. The odor intensity of the three isomers of methyl jasmonate present in a commercial preparation as they eluted from a fused silica bonded and cross-linked OV101 column, 1 uL. injected.   B. The odor intensity of the two Z- epimers as they elute from a 5 micron silica gel column in 15% EtOAc/-hexane.

Another example occurred during some recent studies involving methyl jasmonate. Many female lepidoptera produce male attractant pheromones, but in one species, Grapholitha molesta (Busck), the male produces an herbal smelling pheromone as well (19). The chance discovery that the herbal smelling compound was also present in lemon peels facilitated its characterization as (+)-methyl epijasmonate. During this work the three dominant isomers of methyl jasmonate were separated and their sensory properties examined as they eluted from a high resolution gas chromatographic column (Figure 2).

This convinced us that the odor activities of the two epimers of (Z)-methyl jasmonate were the same. However, when we smelled the solutions of the two epimers purified by high pressure liquid chromatography, the results were quite different. The pure methyl epijasmonate solution had a much stronger odor than the 17-fold more concentrated solution of methyl jasmonate. This conflict was resolved only when very dilute samples of methyl jasmonate were analyzed using a new method for the sensory analysis of gas chromatographic effluents, called CHARM.

CHARM Analysis

CHARM, an acronym from Combined Hedonic Response Measurement, uses high resolution gas chromatography, effluent sniffing and computerized data collection (20). The configuration of the sniffer has been described elsewhere (6), and the capillary columns were attached directly to it without the use of a splitter. This minimized the loss of resolution from dead volumes, and disturbances in retention time due to back pressure fluctuations. The sniffer-column interface was constructed from an FID detector and was designed to maintain a

constant back pressure rather than make-up gas flow rate (21).
Constant pressure drop across capillary columns is required
for the production of reproducible retention times.

A subject sits in front of a computer terminal interacting
with a program while simultaneously sniffing the gas chromatographic effluent. The program uses a clock routine to
record the precise time that a key stroke was made by the
subject. When an odor is detected by the subject, he strikes
the spacebar followed by a key coded for the odor quality
perceived by the subject. Finally, the subject strikes a key
the moment the odor changes or disappears, and then repeats
the process when he perceives the next odor. Throughout the
entire run the program displays prompts to keep the subject
aware of his present options.

When the gas chromatographic run is completed the mean time
after injection, duration and quality attribute of each sensory event is stored in a table. The output end of the column
is then moved to an adjacent FID detector of similar configuration and a series of normal hydrocarbons are separated
under the same chromatographic conditions used for the sensory
analysis. Their retention times are used to convert the data
in the sensory table to a retention index basis. This produces data which can be reproduced and scrutinized at other
places and times. A typical result would be: a lemon-like
odor can be separated from an extract of a male moth, with a
retention index of 1641 on methyl silicone.

The next step is to repeat the analysis of the same sample in
two different ways. One way is to have several different
subjects examine the sample or have the same subject repeat
the analysis. Then the resulting tables can be combined and
plotted to produce a gas chromatogram in which the ordinate is
the number of coincident responses. Figure 3 shows a

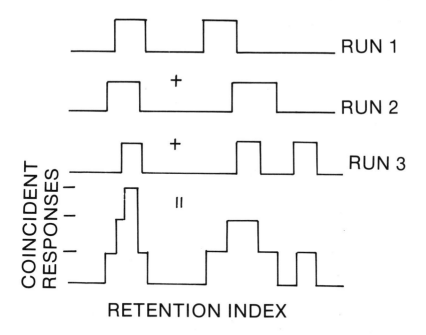

Figure 3   A diagrammatic representation showing the combination of single sensory events to produce a charm-response gas
chromatogram.

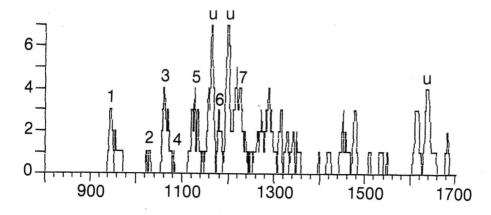

Figure 4   A charm-response gas chromatogram of a Freon 113
extract of apple juice using 4 subjects and 4 concentrations.
Numbered peaks correspond to known apple volatiles: 1, methyl
butanoate; 2, ethyl butanoate; 3, ethyl 2-methylbutanoate; 4,
butyl acetate; 5, isopentyl acetate; 6, pentyl acetate; 7,
2-(E)-hexenal.   "u" are unknown compounds.

schematic diagram of the process. The exact meaning of the
resulting peak areas will be discussed elsewhere (20) but the
resulting chromatograms are called coincident response gas
chromatograms.

The second way to analyze a sample is to dilute the sample,
run it again, follow by another dilution, and so on. If a
series of dilutions of an extract of a complex natural product, for example apple juice, are analyzed using CHARM then
the combination of the resulting sensory tables produce a
chromatogram like the one in Figure 4. The ordinate of the
graph in this case is the combination of the sensory response
tables from four different subjects and four dilutions of the
sample. Again, the precise meaning of the peak areas is not
simple, but the peak heights are certainly related to the odor
intensity of the compounds eluting from the gas chromatograph.
Using mass spectroscopy, 7 of the peaks in this charm-response
gas chromatogram could be ascribed to compounds known to be in
apples, but the two tallest peaks are unknown. The peak at
1640 is probably due to beta damascenone.

If the samples are diluted until no odor is detected in the
gas chromatographic effluent, the resulting peak heights
relate directly to the gas phase detection thresholds for the
eluting compounds. This requires an account of the dilutions
involved in the experiments. Such an account was taken in the
charm-response gas chromatogram shown in figure 5. This
figure shows a comparison between the FID response observed
during the chromatography of a sample made from equal aliquots
of extracts from 40 different apple cultivars, and the combination of the 40 charm-response gas chromatograms obtained
for these cultivars. This remarkable chromatogram contains
some 3000 sensory events and provides a unique view of the
sensory characteristics apples in general (22). The study of
the precise chemical causes for this charm-response gas chromatogram continues in this laboratory.

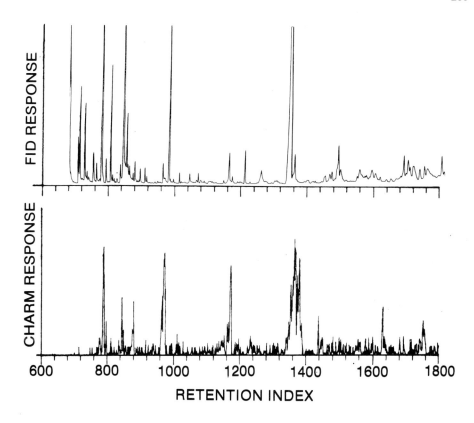

Figure 5 A comparison between the FID response and the charm response for the chromatography of a combined sample of 40 cultivars of apples on OV101 (22).

The Odor Thresholds of Methyl Jasmonate Isomers

Let us return to the question of the odor-activity of the isomers of methyl jasmonate and the use of CHARM to answer it. Using high pressure liquid chromatography, very pure samples of the two epimers were prepared, and their odor thresholds (detection) were determined in water. (±)-Methyl jasmonate had an odor threshold of 5700 ng/mL while (±)-methyl epijasmonate had a threshold of only 13 ng/mL (23). It is conceivable that some conversion between the two isomers, as shown in Figure 6, could have occurred between the time the epimers were purified and the sensory analyses were conducted.

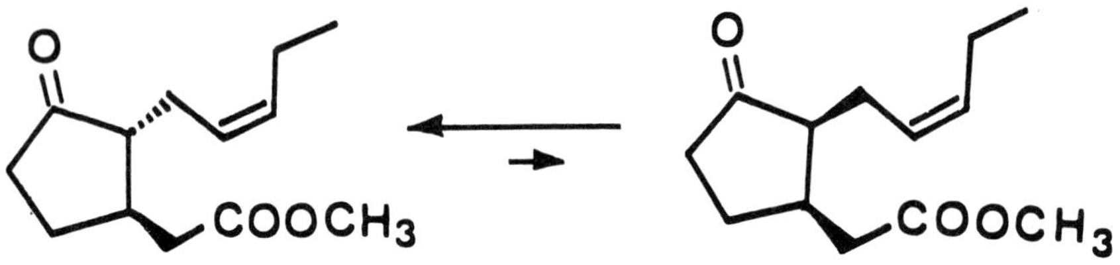

Figure 6   The enolate mediated equilibrium between methyl jasmonate and the thermodynamically less stable methyl epijasmonate.

A contamination of only .2% of methyl epijasmonate in methyl jasmonate would account for the threshold determined for this compound. In a CHARM analysis however, the epimers are examined at the instant they are separated.

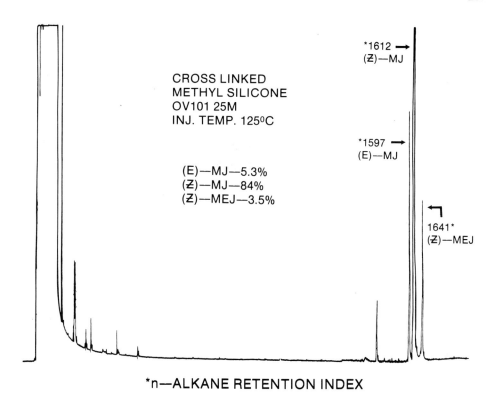

CROSS LINKED
METHYL SILICONE
OV101 25M
INJ. TEMP. 125ºC

(E)—MJ—5.3%
(Ƶ)—MJ—84%
(Ƶ)—MEJ—3.5%

*1612 →
(Ƶ)—MJ

*1597 →
(E)—MJ

↰
1641*
(Ƶ)—MEJ

*n—ALKANE RETENTION INDEX

Figure 7   The gas chromatographic separation of a commercial
sample of methyl jasmonate on crosslinked, bonded-phase methyl
silicone at a film thickness of .51um.   The column was 25m X
.31mm fused silica.

Figure 7 shows a gas chromatographic separation of the isomers
of methyl jasmonate contained in a commercial preparation.
This sample contained 5% (E)-methyl jasmonate as well as the
two epimers of the Z form near their equilibrium concentrations.

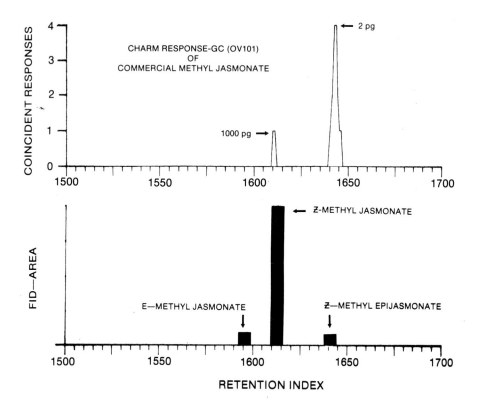

Figure 8   A charm-response gas chromatogram produced by the successive threefold dilution of this commercial methyl jasmonate sample until no odor could be detected eluting from the gas chromatograph.   A bar graph based on the FID peak areas is shown below for comparison.

Associated with each peak is the smallest amount of the compound detectable during the CHARM analysis.   Clearly the epi isomer has the lowest gas phase threshold, but 1000 picograms of methyl jasmonate could still be detected in the gas chromatographic effluent.

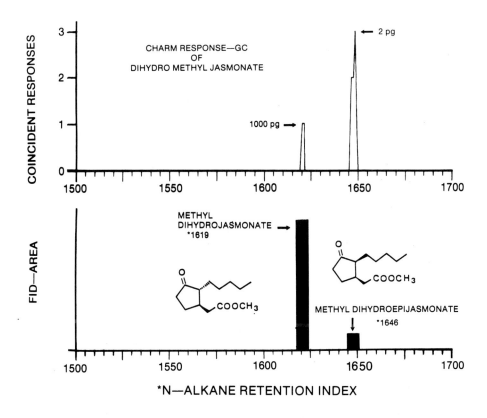

Figure 9   The charm-response and FID gas chromatograms of hydrogenated methyl jasmonate.

The presence of E-methyl jasmonate in the commercial sample indicates that some E-methyl epijasmonate may be eluting at the same retention index as Z-methyl jasmonate and producing the odor observed at that elution time.   To resolve this problem, the commmercial sample was reduced with hydrogen gas over palladium on charcoal.   The charm-response gas chromatogram of the resulting methyl dihydrojasmonates shown in Figure 9 is identical to that shown in Figure 8 except for different retention indices.

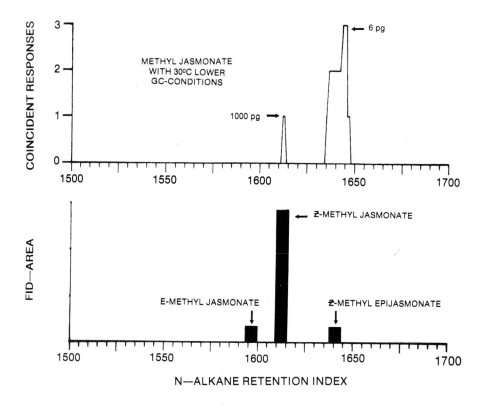

Figure 10   The charm-response and FID gas chromatograms on OV101 at an elution temperature of 170 C.

An additional possibility is that the temperature at the end of the gas chromatographic run stimulates the epimerization reaction. Figure 10 shows that 1000 pg of methyl jasmonate is still detectable even though the maximum temperature was 30 C lower the run in Fig. 8. It seems therefore that methyl jasmonate does have an odor although its gas phase threshold is 500 times lower than that of its epimer.

Conclusion

The rigorous application of sensory analysis of gas chromatographic effluents should produce meaningful bioassays. If
the following considerations are accepted as necessary requirements of the bioassay: 1, the procedure preserves the
integrity of the gas chromatographic separations at the human
interface; 2, compounds are examined near their detection
thresholds; 3, retention indexing is used; 4, the use of
psychological estimations of intensity are minimized; the data
in Table 1 support the conclusion that the CHARM procedure
provides a rigorous technique.

Table 1.   Odor Threshold Ratio of
Methyl Jasmonate/Epijasmonate.

Sensory Panel	440
Charm Response (200 C)	500
Dihydro Derivative	500

The sensory panel data is a proper measure of the relative
threshold of these two compounds dispersed in water and the
remarkably similar result derived from charm-response is a
measure of the relative gas phase threshold for these same
compounds in a gas chromatographic effluent. Not only is the
CHARM technique an appropriate bioassay for gas chromatographic effluents and a simple procedure for the determination
of gas phase detection and recognition thresholds, it can in
certain situations be a meaningful alternative to sensory
panels.

266

References

1) Teranishi, R.; Issenberg, P.; Hornstein, I.; Wick, E.L.
   Flavor Research-Principles and Techniques." 1971, p. 17,
   Marcel Dekker, New York.

2) James, A.T.; Martin, A.J.P. Analyst. 1952, 77, 915-.

3) Fuller, G.H.; Steltenkamp, R.; Tisserand, G.A. The Gas
   Chromatograph with Human Sensor: Perfumer Model. Annals.
   New York Academy of Sciences. 1964, 116, 711-724.

4) Potter, R.H.; Daye, J. Apparatus to Introduce Moisture
   into Effluent Gas. The Givaudan Flavorist. 1970, 2, 8-.

5) Dravnieks, A.; O'Donnell, A. Principles and Some Techniques of High-Resolution Headspace Analysis. J. Agric.
   Food Chem. 1971, 19, 1049-1056.

6) Acree, T.E.; Butts, R.M.; Nelson, R.R.; Lee, C.Y. Sniffer
   to Determine the Odor of Gas Chromatographic Effluents.
   Anal. Chem. 1976, 48, 1821-1822.

7) Rapp, A.; Knipser, W.; Engel, L.; Ullemeyer, H.; Heimann,
   W. Fremdkomponenten im Aroma von Trauben Weinen interspe-
   zifischer Rebsorten. I. Die Erdbeernote. Vitis. 1980, 19,
   13-23.

8) Pyysalo, T.; Honkanen, E.; Hirvi, T. Volatile Constituents and Odor Quality of Apple Juice. Int. Fed. of Fruit
   Juice Producers. XVI. Symposium. 1980, 343-350.

9) Casimir, D.J.; Whitfield, F.B. Flavour Impact Values: A
   New Concept for Assigning Numerical Values for the Po-
   tency of Individual Flavour Components and Their Contri-
   bution to the Overall Flavour Profile. Int. Fed. of Fruit
   Juice Producers. XV.Symposium: Flavors of Fruits and
   Fruit Juices. 1978, 325-345.

10) Guadagni, D.G.; Okamo, S.; Buttery, R.G.; Burr, H.K.
    Correlation of Sensory and Gas-Liquid Chromatographic
    Measurement of Apple Volatiles. Food Technol. 1966, 20,
    518-521.

11) Von Sydow, E.; Andersson, K.A.; Karlsson, G.; Land, D.;
    Griffiths, N. The Aroma of Bilberries (Vaccinium myrtil-
    lus L.) II. Evaluation of the Press Juice by Sensory

Methods and by Gas Chromatography and Mass Spectrometry.
Lebensm.-Wiss. u.-Technol. 1970, 3, 11-17.

12) Noble, A.C.; Flath, R.A.; Forrey, R.R. Wine Headspace
Analysis. Reproducibility and Application to Varietal
Classification. J. Agric. Food Chem. 1980, 28, 346-353.

13) Tassan, C.G.; Russel, G.F. Sensory and Gas Chromatographic Profiles of Coffee Beverage Headspace Volatiles
Entrained on Porous Polymers. J. Food Sci. 1974, 39, 64-
68.

14) Acree, T.E. Flavor Characterization. Kirk Othmer
"Encyclo. Chem. Technol." Volume 10. Third Edition. p.
444, 1980, John Wiley and Sons Inc.

15) Cain, W.S. Science, 1979, 203, 467-.

16) Schreyen, L.; Dirinck, P.; VanWassenhove, F.; Schamp, N.
Volatile Flavor Components of Leek. J. Agric. Food Chem.
1976, 24, 336-341.

17) Selke, E.; Rohwedder, W.K.; Dutton, H.J. A Micromethod to
Generate and Collect Odor Constituents from Heated Cook-
ing Oils. J.A.O.C.S. 1972, 49, 636-640.

18) Nursten, H.E.; Woolfe, M.L. An Examination of the Vola-
tile Compounds Present in Cooked Bramley's Seedling
Apples and the Changes they Undergo on Processing. J.
Sci. Food Agric. 1972, 23, 803-822.

19) Baker, T.C.; Nishida, R.; Roelofs, W.L. Close-Range
Attraction of Female Oriental Fruit Moths to Herbal Scent
of Male Hairpencils. Science. 1981, 214, 1359-1361.

20) Acree, T.E.; Cunningham, D.G.; Barnard, J. Manuscript in
Preparation.

21) Butts, R.M.; Braell, P.A.; Cunningham, D.G.; Acree, T.E.
Publication submitted to Anal. Chem., 1984.

22) Cunningham, D.G. A Description of the Odor-Active Volatile Components of Apple Cultivars. PhD Thesis, 1984,
Cornell University, Ithaca.

23) Acree, T.E.; Nishida, R.; Fukami, H. The Odor Detection
Threshold of the Stereo Isomers of Methyl Jasmonate in
Water. Submitted to J. Agric. Food Chem. 1984.

SIGNIFICANCE OF THE SNIFFING-TECHNIQUE FOR THE DETERMINATION OF ODOUR THRESHOLDS AND DETECTION OF AROMA IMPACTS OF TRACE VOLATILES

F. Drawert and N. Christoph

Institut für Lebensmitteltechnologie und Analytische Chemie
der TU München, Freising-Weihenstephan, FRG

Introduction

The first two periods of flavour research were characterized
by instrumental analysis and differentiation of primary and
secondary aroma compounds in food. Development of high resolution capillary columns in combination with mass spectrometry
allowed separation and identification of hundreds of components. Since the last decade, a third period of flavour research is characterized by questions like the significance of
sensory, olfactory and physiological properties of aroma compounds, their structure-activity relationship in chemoreception, and for example the significance of stereo-specific trace
volatiles in food aromas. Computer assisted analyses help to
interprete and correlate analytical and sensory data with statistical operations. Especially the search for "character impact compounds" and volatiles with high aroma values is forced,
in order to characterize odour quality of a complex aroma by
means of some few important constituents of typical odour
qualities.

The gaschromatographic sniffing test for the determination of olfactory properties of aroma compounds

A well-known method in flavour research is the gas chromato-

graphic sniffing test, which is used to characterize the odours
of single constituents of a complex mixture of aroma compounds
emerging from the sniffing port. Two examples (fig. 1 and 2)
illustrate that this method is important to detect aroma compounds in trace amounts, which can be very effective in the
total aroma.

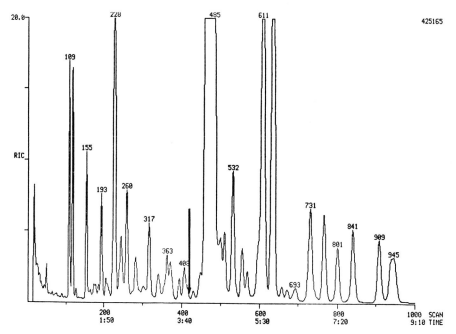

Figure 1   Part of a gas chromatogram of a pineapple aroma
           extract

Figure 1 shows the part of a gas chromatogram of a pineapple
aroma extract. The arrow marks a small area in the chromatogram, where by sniffing the effluent, a typical pineapple
odour could be recognized, but the FID sensitivity was too low
to show a peak. By enrichment, allyl hexanoate was identified,
which was never found before to be a natural flavour constituent of pineapple (1).

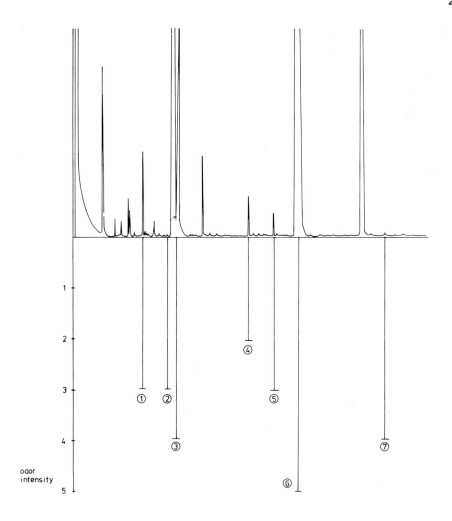

Figure 2   Part of a sniffing chromatogram of an apple aroma
concentrate. Classification of odour intensities on a 0-5 scale;
Odour qualities:
1 = isobutylacetate : fruity   5 = cis-2-hexenal : like grass
2 =   ?   : black currant   6 = trans-2-hexenal : green
3 = 1-butylacetate : fruity, pear   7 =   ?   : mushroom
4 = isopentylacetate: fruity, apple

Fig. 2 shows the sniffing chromatogramm of an apple aroma con-

centrate. Two very small peaks (peak number 2 and 7), not yet

identified, have relative high odour intensities, but cannot

be attributed to apple aroma. Thus analyzing isolates and
concentrates by means of the gas chromatographic sniffing tests,
information about the olfactory significance of their single
constituents can be obtained. However, to characterize an aroma,
it is not satisfying to analyze solvent extracts of food material
only. Olfactory senses base their judgement of aroma on concentrations of the respective volatiles in the headspace of
the food material. A complete characterization of an aroma can
only be achieved, when chemical as well as olfactory properties
of its constituents in the headspace are qualitatively and
quantitatively known too.

By development of synthetic porous polymers like Tenax GC and
Porapak Q, it is possible to enrich headspace volumes of 10 1
or more, so that volatiles with headspace concentrations of
less than one ng/l can be identified. A greater problem is the
lack of reliable quantitative olfactory data of most aroma
compounds. One important olfactory property of an odour substance is its odour threshold value. Thresholds of many odour
and aroma compounds, determined in air and water, have been already published or compiled in literature. However, these values scatter so much that objective statements for example of
aroma significance are often impossible. Van Gemert and Nettenbreijer (2) compiled odour threshold values of about 2000
data in air and water. In order to show the wide ranges of
threshold values in air for the same compounds, the compiled
detection threshold values of homologous n-alcanols are visua-
lized in figure 3.
Especially olfactometric threshold determinations lack of reproducibility. The reasons are different methodological parameters like olfactometers with different methods of stimulus
presentation or concentration adjustment,and different psychophysical variables like the number of test subjects, the degree of experience of the subjects with the stimulus and with
the test procedure, verification with replicate testing etc.
(2,3). Determining odour thresholds, purity of the test sub-

stance is an important factor too. However many threshold values have been determined 50 and more years ago (2), prior to
The use of gas chromatographic analysis; these values have only
a very limited realibility.

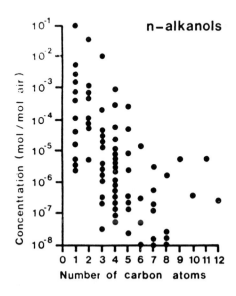

Figure 3   Compilation of detection odour threshold values of
homologous n-alcanols in air, by van Gemert L.J. and Nettenbreijer, A.H., 1977 (2)

Thus   for a combined interpretation of quantitative headspace
analysis we aimed to develop a reproducible and standardized
method to determine odour threshold values of single aroma compounds in the effluent of a sniffing gas chromatograph.
Quantitative olfactometric data of some few compounds in gas
chromatographic effluents have been already published (4,5,6),
but   none . study lists threshold values of a greater number of
aroma compounds by means of a standardized method.

Development of a method for determination of odour recognition threshold values in gaschromatographic effluents.

Some important apparative and methodical prerequisites and standardizations of the GC sniffing test were necessary to determine odour thresholds of single odour stimuli eluting from the sniffing port.

Apparatus

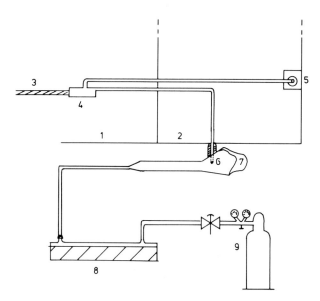

Figure 4   Arrangement of the sniffing port

1 = column oven
2 = detector oven
3 = 5 m glass column (i.d.2mm),filled with FFAP on Varaport 30
4 = constant stainless steel splitter,split ratio(120°C) 1:5.5
5 = FID
6 = sniffing port, stainless steel tube (i.d. 1 mm)
7 = stainless steel olfactometer mask
8 = arrangement for humidifying effluent
9 = synthetic air supply (flow rate: 100 ml/min.)

First of all, we had to arrange that the eluted stimulus reaches the biodetector epithelium, like the FID, in the same qualitative and quantitative reproducible manner. Figure 4 shows the

arrangement of the sniffing port. The stainless steel tubes
from the constant splitter to the sniffing port and to the FID
are equally long. The tube leading into an olfactometer mask
is heated up to 220°C, except for the last two centimeters,
which are isolated with teflon. So without time delay the
stimulus can emerge in the centre of the olfactometer mask,
parallel to the FID signal.

The olfactometer mask fits to the nose, resulting in a constant position of the nostrils about 2 cm over the sniffing
port. In this way the stimulus can be inhaled directly, without
irritating effects of hot sites and without losses and irrita-
tion by ambient air. Directing the emerging stimulus to the
nostrils, the gentle stream of synthetic humidified air (fig.4)
prevents condensation of the stimulus on the walls of the mask
and the drying up of the olfactory mucosa.

Qualitative and quantitative characterization of the stimulus in the gaschromatographic effluent

A further prerequisite was the qualitative and quantitative
characterization of the eluted stimulus. First the purity of
each test substance was examined. During gas chromatographic
elution it is generally purified, so that olfactory efficient
impurities do not disturb the characterization of the main
stimulus. By the following equation stimulus amount SA (ng)
eluted from the sniffing port can theoretically be calculated
from the concentration c (ng/µl) of the solution of the test
substance, the volume V (µl) injected with a gas-tight syringe,
the purity p (%) of the substance, and the split ratio determined to the corresponding column temperature:

$$SA = (c \cdot V \cdot p) - (c \cdot V \cdot p) / r$$

Enriching about twenty stimuli on Porapak Q tubes, connected
with the sniffing port, the amount of eluted substance can be

determined by analyzing the solvent extract of the adsorption tube. Thus, losses in the gas chromatographic system, artifacts or decompositions can be detected.

For comparable investigations of olfactory properties it is very important to define standardized stimulus parameters. The stimulus eluting from the sniffing port represents a so-called pulse-type stimulus (7) in contrary to the so-called constant stimulus, delivered from generally used "constant dynamic ol-factometers".

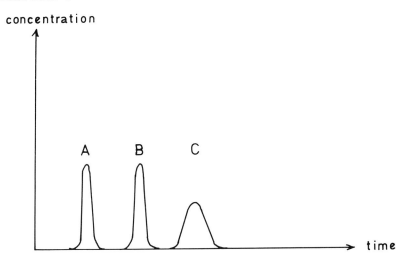

Figure 5   Time profiles of pulse-type stimuli of three diffe-rent odour substances, figuring the same stimulus amount.

Figure 5 schematically shows the same amounts of pulse-type sti-muli of three different components A, B, and C. Objective mea-surements of comparable olfactory properties like intensity are only possible by comparing stimulus A and B, which have identical time profiles. Comparing stimulus C with A or B is not possible, as with C a smaller number of odour molecules is eluted in the same specific time unit.

The time profile of a stimulus, eluted from the sniffing port, can be described by its elution time, calculated from the peak width of the parallel appearing FID signal and the chart speed.

Table 1 illustrates the influence of different elution times
on intensity sensation of the same stimulus amount.

Table 1   Influence of stimulus elution time and concentration in the carrier gas on perceived intensity of 5.5 ng n-
nonanol, eluted from the sniffing port; classification of
odour intensities on a 0 - 5 scale
t = stimulus elution time (sec.)
c = stimulus concentration in the carrier gas (ng/ml)
h = peak height of the FID signal (mm)

t	c	h	odour intensity	
11	1.15	9	very weak	(1)
9	1.3	13	weak	(2)
7	2.0	15	clear	(3)
5	3.0	19	clear	(3)

If elution time of a stimulus of 5.5 ng n-nonanol is shortened
from 11 to 5 seconds, the concentration of the stimulus in
carrier gas increases and perceived intensity changes from
very weak to clear. Thus during gas chromatographic sniffing
test of a complex mixture of aroma compounds, different peak
widths, which can not be prevented even by use of capillary
columns or temperature programming, may lead to misinterpretations of relative intensities of the single components.

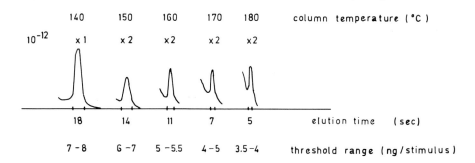

Figure 6   Influence of the elution time of n-octanol stimuli,
calculated from the peak width of the parallel appearing FID
signal and the chart speed, on their recognition thresholds.

Figure 6 shows as an example the influence of the elution
time on the smallest just recognizable amount of n-octanol
stimuli.
It decreased from seven to about 3.5 nanograms, when elution
time was changed from 18 to 5 seconds by raising GC column
temperature from 140 to 180°C.
We defined the recognition threshold of a pulse-type stimulus
as its total amount, which can just be recognized during elution within 5 seconds, corresponding to the peak width of the
parallel appearing FID signal. On the basis of this constant
elution time the total stimulus amount can be inhaled within
one slow and steady breath of about 4 seconds, independently
from the individual respiration volume. The threshold depends
only on the limited number of eluted odour molecules, which
can reach the receptor site within 4 seconds. It is a measure
for the detectability of an aroma compound, as well as a measure for sensitivity of the olfactory system (7).

Reproducibility of recognition threshold value determinations

6 to 8 judges were used to examine the reproducibility of the
method by determining recognition thresholds of some homologous n-alkanols. In preliminary experiments we determined the
approximate threshold ranges and the particular gas chromatographic conditions for stimulus elution times of 5 seconds.
After elution of the solvent peak the judge was told to start
with slow and steady breathing and to indicate immediately the
recognition of the stimulus. Inserting blank and repetition
stimuli, about 15 single stimuli in the expected threshold
range were presented to each judge, until the stimulus was no
longer recognized. The judge could not watch the FID signal,
so that an objective control of his subjective response was
realized and responses by chance were excluded.
Table 2 shows the results of the threshold determinations.
Every judge just recognized the single alkanols in such
amounts that a narrow range for more than 50 % of the judges

Table 2 Determination of recognition threshold ranges of n-alkanol stimuli by several judges·; stimulus elution time corresponding to the peak width : 5 sec. recognition threshold ranges : nanogram / stimulus

$th_{>50}$ = threshold range, just recognized by more than 50 % of the judges

judge (no.)	C_4	C_5	C_6	C_7	C_8	C_9
1	100 - 120	80 - 100	40 - 48	43 - 52	9 - 11	6.5 - 8
2	-	-	36 - 44	27 - 33	6 - 8	5 - 6
3	90 - 110	90 - 110	55 - 65	32 - 38	5 - 6	-
4	80 - 100	90 - 110	27 - 33	33 - 41	6 - 8	3 - 4
5	70 - 90	80 - 100	35 - 42	20 - 25	5 - 7	2.5 - 3
6	110 - 130	90 - 110	55 - 65	31 - 37	8 - 10	5 - 6
7	70 - 90	-	-	38 - 46	5 - 7	-
8	110 - 130	100 - 120	35 - 42	47 - 57	-	-
$th_{>50}$	70 - 100	80 - 110	35 - 45	30 - 45	5 - 8	4 - 6

could be ascertained. With regard to all prerequisites, determinations of recognition threshold values of stimuli of limited stimulus amount and duration seem to be practicable with
every gas chromatograph equipped with an effluent splitter.

Recognition threshold values, odour qualities and headspace concentrations of lemon aroma compounds

The aim of further investigations was the determination of
olfactory significant aroma compounds of lemon on the basis
of odour qualities and recognition thresholds of their stimuli
eluted from the sniffing port, and their concentrations in
headspace. Threshold ranges of 73 aroma compounds mostly
identified by own mass spectral analysis of lemon aroma were
determined (8). Although these threshold values were not veri-
ficated by all judges, determining the thresholds of n-alkanols, they were confirmed by a sufficient number of test
series.

Table 3 shows some of 60 identified lemon aroma compounds
arranged with increasing threshold values. The aroma compounds
with typical lemon, respectively citrus-fruity odour, and low
thresholds are octanal, nonanal, decanal, neral,and geranial.
Other compounds with low thresholds, which first cannot be
attributed to lemon aroma are for example undecanal, dodecanal,
1,8-cineole,and (+)-linalool. The thresholds of terpene hydrocarbons, except myrcene,as well as those of some terpene alcohols and -esters like nerol, α-terpineol, neryl- and geranyl-
acetate, are relatively high.

The headspace was enriched on Porapak Q and Tenax GC tubes
using three different sampling methods : the sweeping technique
with nitrogen and the drawing off by a continuous and by a
discontinuous suction method. Adsorption and desorption efficiency of adsorbing tubes and calibration factors were determined by means of model systems. Sampling flow rate was varied
in the range of 10 to 30 ml/min adsorbing headspace volumes
of one to about three liters, while sampling temperature of

Table 3   Recognition threshold ranges and odour qualities of
stimuli of some lemon aroma compounds, eluted from the sniffing port
stimulus elution time corresponding to the peak width : 5 sec.
threshold ranges : nanogram / stimulus
        N   : number of single threshold determinations

aroma compound	threshold range	N	odour quality
1-octanal	0.5 - 0.9	50	lemon-fruity,aldehyde
ethylhexanoate	0.6 - 1.0	20	fruity, rum-like
1-undecanal	0.9 - 1.2	26	aldehyde
1-dodecanal	1.4 - 2.0	27	aldehyde
1-hexanal	2 - 4	50	green
1-decanal	3 - 4	23	orange-fruity,aldehyde
(+)-linalool	3.5 - 4.0	15	flowery, with fruity note
neral	3 - 5	20	lemon
1,8 cineole	3 - 5	20	medicinal,eucalyptus
1-nonanal	4.5 - 6.0	26	lemon-fruity,aldehyde
geranial	5.5 - 7.0	20	lemon
citronellal	5.0 - 6.5	15	aldehyde, soapy
myrcene	13 - 15	15	like hops
geraniol	15 - 20	32	flowery
citronellol	80 - 100	25	flowery
limonene	220 - 230	41	terpeny, fruity-sweet
nerol	320 - 370	37	flowery,with fruity note
geranylacetate	1200 - 1800	16	fruity
nerylacetate	2000 - 2500	15	citrus-fruity
α-pinene	2500 - 2900	22	terpeny,turpentine-like
ß-pinene	3500 - 3800	45	coniferous
α-terpineol	10000 - 11000	15	flowery

25°C and equilibration time of 30 minutes were kept constant.
After elution of adsorbed volatiles with ether and addition
of internal standard substances, the eluate was concentrated
and separated in an unpolar and a polar fraction by column
chromatography on silica gel. The compounds were identified by
mass spectral analysis and headspace concentrations were
determined by analyzing concentrates on a 25 m CW 20 M glass
capillary column (8).
Thus with a detection sensitivity of about 100 ng/l a headspace concentration range, respectively a mean value for each
component in headspace could be ascertained, representing the
approximate quantitative proportions as perceived by the nose.
According to the morphological differentiation of lemon and
other citrus fruits, peel and fruit, as well as juice were
separated prior to the headspace analysis. Determining influence of interactions of volatiles with different lemon tissues
on headspace concentrations, a cold pressed lemon peel oil
(Aromachemie, Aufsess) and a water-peel oil emulsion of the
peels, separated by pressing, were additionally analyzed.
Table 4 shows headspace concentration ranges and mean values
of some selected aroma compounds in different lemon samples.

The terpene hydrocarbons, especially limonene are the main
constituents in the headspace of all samples. The main polar
compounds in cold pressed lemon peel oil and the water-peel
oil emulsion, separated from the peel are citral, 1- octanal
and 1-nonanal, while nerol, geraniol and n-alcanols are dominating in the headspace of lemon peel pieces. Concentrations
of peel oil aroma compounds in the headspace of peel-oil-free
lemon juice are considerably minor, while compounds of high
volatility, not yet identified, are dominating (8). These
results illustrate the well-known influence of the sample
matrix on the headspace concentrations of the volatiles (9,10).

283

Table 4 Concentration ranges of some volatiles in the headspace of lemon samples
(in 10^{-6} g/l) 1 = mean values

compound	coldpressed ital. lemon peel oil	small pieces of span. lemon peels	water-peel oil emulsion of span. peels [1]	peel-oil-free lemon juice[1]
α-pinene	450 – 750	600 – 760	280	11
β-pinene	1540 – 2600	1750 – 2450	1160	50
myrcene	140 – 225	204 – 240	162	10
limonene	5000 – 7900	6030 – 7380	5920	300
1,8-cineole	6 – 10	10 – 12	10	0.1
1-octanal	7 – 10	0.2 – 0.8	5.2	0.1
1-nonanal	3.4 – 4.8	0.4 – 0.7	4.9	0.2
citronellal	1.9 – 3.2	0.3 – 0.9	2.7	0.2
neral	11.1 – 15.5	2.0 – 3.4	10.1	0.3
geranial	12.7 – 18.5	2.4 – 3.0	12.3	0.6
nerylacetate	1.0 – 2.0	0.5 – 0.9	1.2	0.1
geranylacetate	1.5 – 2.8	0.4 – 0.5	0.7	0.1
1-hexanol		1 – 3	0.1	–
1-octanol		0.5 – 0.8	–	–
1-nonanol		0.5 – 9.9	–	–
(+)-linalool	6.5 – 8.6	3.6 – 4.7	3.3	0.1
nerol	0.5 – 0.7	5.1 – 7.6	0.7	0.1
geraniol	0.3 – 0.4	4.2 – 5.1	0.5	0.3
citronellol		0.6 – 1.3	–	–
α-terpineol	1.9 – 2.8	2.0 – 2.8	1.8	0.2
terpineol-4	0.8 – 1.1	0.6 – 0.8	0.9	0.2
2-methyl-2-hepten-6-one	7.2 – 9.3	0.1 – 0.2	0.1	0.1

Combined interpretations of analytical and olfactometric data

Calculations of relative aroma values in the headspace

To quantify aroma effectivity of single lemon volatiles in the headspace, the concept of "aroma value" respectively "odour unit" applied by Rothe (11) and by Guadagni (12) was transferred to threshold values and headspace concentrations of lemon volatiles. As the threshold value of a pulse-type stimulus is not related to a defined dilution volume, but represents the smallest, just recognizable stimulus amount, determined by standard conditions, the calculation of aroma values was based on an arbitrary respiration volume of 100 ml per stimulus. Thus from the average concentration c (ng/100 ml headspace) and the average recognition threshold value s (ng/stimulus) of an aroma compound, its relative aroma value in the headspace (A_H) was defined by the quotient:

$$A_H = \frac{c}{s}$$

Table 5 shows lemon volatiles with the highest relative aroma values in the headspace of the different lemon samples, except the peel-oil-free lemon juice (tab. 4).

Table 5 Lemon aroma compounds and their highest relative aroma values A_H in the headspace of different peel-oil samples

aroma compound	A_H
limonene	2900
myrcene	1500
γ-terpinene	400
1-octanal	1200
neral	350
geranial	280
1,8-cineole	250
(+)-linalool	220
ethylhexanoate	170
1-nonanal	100
geraniol	30
1-octanol	30
1-nonanol	30

Relative aroma values in table 5 are arranged on an arbitrary
scale, for which the base unit is the particular recognition
threshold value of a standardized pulse-type stimulus. According to Steven's law, they are not a psychophysical measure of
intensity in the suprathreshold range, as perceived intensity
is not linearly proportional to the concentration (13).
Terpene hydrocarbons are an example that the concept of aroma
value can lead to some misinterpretations. They generally have
no typical fruity, lemon-like odours (14), and intensity sensation in suprathreshold range was judged to be relatively
weak. The main reason for the high aroma values of limonene,
myrcene and γ-terpinene are their very high headspace concentrations to be found in all citrus essential oils (15). Thus
discussing aroma values with regard to aroma effectivity,
odour quality and intensity in the suprathreshold range are
important parameters. However, other psychophysical phenomena
like additive, synergistic, and antagonistic interactions of
volatiles also have to be considered (3, 13, 16).

In figure 7 a direct relationship can be recognized between the
aroma values and odour qualities of certain polar lemon volatiles and the total aroma of different lemon samples. Evaluating the total aroma, lemon peel pieces additionally had a
typical flowery note, whereas the other samples, described by
lemon-fruity odour, had not. We assumed additive sensations of
fruity, lemon-like odours resulting from the aldehydes 1-octa-
nal, 1-nonanal and citral (17), and of flowery odour qualities
resulting from (+)-linalool, nerol, geraniol, and citronellol.
The headspace of lemon peel pieces is characterized by a well
balanced ratio of aroma values of volatiles with fruity and
flowery odours, (fig 7) for which the reasons are the relative
high headspace concentrations of the alcohols and the relative
low concentrations of aldehydes. Thus the flowery odour note
can contribute more to the total aroma of lemon peel pieces
than to the other samples, where according to the higher headspace concentrations, aroma values of aldehydes are dominating.

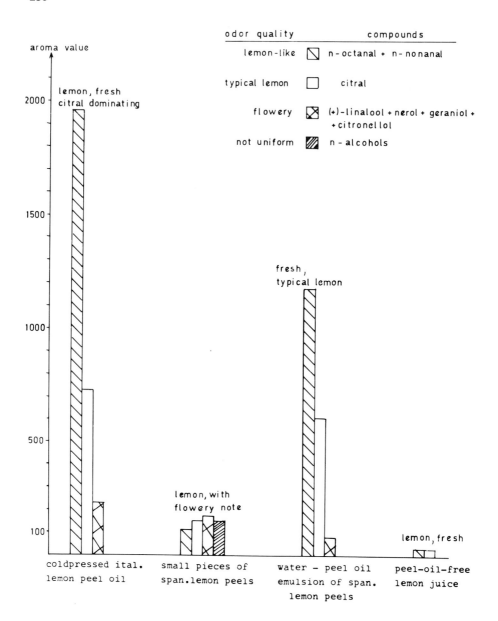

Figure 7 Relationship between headspace aroma values and odour qualities of single lemon volatiles and the total aroma quality of different lemon samples.

This example evidently shows the influence of headspace concentrations on the aroma significance of single components.
Factors influencing volatility like a complex matrix or
different phase states (9,10) can only be included into the
calculations of aroma values based on threshold values in
air and headspace concentrations (11).

Aroma values of aldehydes in the headspace of peel-oil-free
lemon juice are in the range of about 5 to 20 (fig.7).
Evaluating juice headspace concentrate with the gas chromatographic sniffing test, some compounds of high volatility,
i.e. showing short retention times on the FFAP-column (fig.4),
developed fruity odours of relative strong intensity; they
are not yet identified.
For a complete characterization of lemon aroma, these compounds have to be analyzed quantitatively too. However, reproducible analysis of compounds of high volatility by means
of solvent extraction of adsorbing tubes is very troublesome.
Disadvantages of this method are e.g. high adsorption volumes
required, losses during concentrating or the solvent peak.
Thus thermal desorption in combination with column switching
techniques seems to be the best method for quantitative headspace analysis of compounds with high volatility.

Table 6 shows some aroma compounds with very low recognition
threshold values of their pulse-type stimuli. Especially
those of some ethylesters are very low, so for example ethyl
2-methylbutanoate and ethylisopentanoate, with recognition
thresholds of about 10 picogram, eluting from the sniffing
port within 5 seconds.
The ethylesters, also characterized by typical odour qualities
play an important role in many fruit flavours. By calculations
of odour units in water, Guadagni and coworkers (12) identified ethyl 2-methyl-butanoate as an "character impact compound"
of "golden delicious" apple essence. Von Sydow and coworkers
(18) evaluated the same compound to be important for bilberry

aroma. Finally Schultz and coworkers (19)identified this ester
as well as ethylbutanoate and ethylisobutanoate in the headspace of peel-oil-free orange juice. Assuming headspace concentrations of 100 ng / l, aroma values based on these sniffing threshold values would be in the order of 1000 for ethyl
2-methylbutanoate, 200 for ethylisobutanoate,and 50 for ethyl
butanoate. These examples illustrate that trace compounds of
high volatility can have a high aroma effectivity in fruits.

Table 6 Volatiles with low odour recognition threshold ranges
of stimuli eluted from the sniffing port; stimulus elution
time corresponding to the peak width : 5 sec.; threshold ranges : nanogram / stimulus

compound	threshold range	odour quality
ethyl 2-methyl-butanoate	0.006 - 0.012	"golden delicious" apple
ethylisopentanoate	0.007 - 0.01	medicinal
ethylisobutanoate	0.03 - 0.05	fruity,orange-like
ethylpentanoate	0.03 - 0.04	fruity,pineapple-like
ethylbutanoate	0.2 - 0.3	fruity,rancid note
ethanethiol	0.08 - 0.09	cabbage-like
1-octanol	0.5 - 0.9	lemon-fruity, aldehyde

Structure-activity relationships

Finally, odour recognition threshold values determined with
the gas chromatographic sniffing technique can be a basis for
a discussion of structure-activity relationships in human
chemoreception.
In table 7 some examples of different threshold values of
stereo and geometrical isomers are listed like (+)-linalool

and (-)-linalool, nerol and geraniol, and the two isomers of
citral, or the well known difference of thresholds (2) and
odour qualities of α-and β-ionone.
Thresholds of terpene hydrocarbons also distinctly differ.
Bicyclic hydrocarbons, like α-pinene, β-pinene, or camphene
have relative high thresholds, while those of monocyclic
compounds are lower for one order of magnitude. Myrcene, an
acyclic compound, has the lowest odour recognition threshold
of all hydrocarbons examined.

Table 7   Odour recognition threshold ranges of some isomers
and terpene hydrocarbons, determined by gas chromatographic
sniffing technique.
Stimulus elution time, corresponding to the peak width: 5sec.;
threshold ranges: nanogram/stimulus

compound	threshold range	odour quality
(-)-linalool	0.9 - 1.0	flowery
(+)-linalool	3.5 - 4.0	flowery, fruity note
nerol	320 - 370	flowery, fruity note
geraniol	15 - 20	flowery
neral	3 - 5	typical lemon
geranial	5.5 - 7	typical lemon
α-ionone	2 - 4	flowery
β-ionone	450 - 500	flowery, woody note
α-pinene	2500 - 2900	terpeny, turpentine-like
β-pinene	3500 - 3800	coniferous
camphene	2600 - 3000	campherous
γ-terpinene	140 - 160	terpeny, fuel-like
α-terpinene	235 - 245	terpeny
α-phellandrene	290 - 390	sweet, terpeny
myrcene	13 - 15	like hops

290

Conclusions

The gas chromatographic sniffing technique can be used as a
method to determine comparable odour threshold values of single short pulse-type stimuli eluting from the sniffing port
within a constant time unit, e.g. 5 seconds.
These thresholds determined by a standardized method are a
tool to quantify olfactory significance of single volatiles
by combined interpretation of quantitative headspace analysis,
e.g. calculations of aroma values in the headspace of food
materials. They are also a basis for discussing structure-
activity relationship in human chemoreception.

References

1. Nitz, S., Drawert, F.: Chem. Mikrobiol. Technol. Lebensm.
    7, 148 (1982).
2. Van Gemert, L.J., Nettenbreijer, A.H.:"Compilation of
    odour threshold values in air and water", National In-
    stitute for Water Supply, Voorburg, The Netherlands and
    Central Institute for Nutrition and Food Research TNO,
    Zeist, The Netherlands (1977).
3. Pangborn, R.M.: in "Flavour 81", Ed. Schreier, P., 3-32,
    Verlag Walter de Gruyter, Berlin-New York 1981.
4. Fuller, G.H., Steltenkamp R., Tisserand, G.A.: Ann. N.Y.
    Acad. Sci. 116, 711 (1964).
5. Dravenieks, A., O'Donell, A.: J. Agric. Food Chem. 19,
    1049 (1971).
6. Dravenieks, A., Mc Daniel, H.C., Powers, J.J.: J. Agric.
    Food Chem. 27, 336 (1979).
7. Dravenieks, A.: in "Methods in Olfactory Research", Ed.
    Moulton, P.G., Turk, A., Johnston, J.W.jr.: 1 - 61, Aca-
    demic Press, London-New York-San Francisco 1975.
8. Christoph, N.: Diss. TU München, 1983.
9. Land, D.G.: in "Progress in Flavour Research" Eds. Land,
    D.G., Nursten, H.E.: 53-66, Applied Publishers Ltd.,
    London 1979.
10. Radford, T., Kawashima, K., Friedel, P.K., Pope, L.E.,
    Gianturco, M.A.: J. Agric. Food Chem., 22, 1066 (1974).

11. Rothe, M., Wölm, G., Tunger, L., Siebert, H.-J.: Die Nah-
    rung 16, 483 (1972).

12. Guadagni, D.G., Buttery, R.G., Okano, S.: J. Sci. Food
    Agric. 14, 761 (1963).

13. Frijters, J.E.R.: in "Progress in Flavour Research", Eds.
    Land, D.G., Nursten, H.E., 47-51, Applied Publishers Ltd.
    London 1979.

14. Anandaraman, S., Shankaracharya, N.B., Nararajan, C.P.,
    Damodaran, N.P.: Riechstoffe, Aromen, Körperpflegemittel
    2/3, 28 (1976).

15. Shaw, P.E.: J. Agric. Food Chem. 27, 246 (1979).

16. Salo, P.: in "Proc. Symposium of Aroma Research", Zeist,
    The Netherlands, 121-130, Centre for Agric. Public. and
    Doc.(Pudoc) Wageningen 1975.

17. Guadagni, D.G., Buttery, R.G., Okano, S.J., Burr, H.K.:
    Nature (London) 200, 1288 (1963 b).

18. Von Sydow, E., Andersson, J., Anjou, K., Karlsson, G.:
    Lebensm.-Wiss. u. Technol. 3, 11 (1970).

19. Schultz, T.H., Flath, R.A., Mon, T.R.: J. Agric. Food
    Chem. 19, 1060 (1971).

THE USE OF HRGC-FTIR IN TROPICAL FRUIT FLAVOUR ANALYSIS

Peter Schreier and Heinz Idstein
Institut für Pharmazie und Lebensmittelchemie,
Universität Würzburg, 8700 Würzburg, FRG

Werner Herres
Bruker Analytische Meßtechnik, 7500 Karlsruhe, FRG

Introduction

Since the first attempts to use an infrared (IR) interferome-
tric (Fourier Transform or FT) spectrometer as detector combined with a gaschromatograph as separation system (1) a remarkable progress in on-line GC-FTIR technique can be noticed. The
development of instrumental equipment resulted in FTIR instruments fast enough to aquire complete mid-IR spectra within mil-
liseconds. As nowadays capillary columns of low loadability are
increasingly used to obtain higher separation efficiency (HRGC),
this high speed and high sensitivity in IR spectroscopy are needed to achieve several spectral scans over a narrow gas chromatographic peak. Due to the additional IR detection providing
all the advantages of IR information such as characterizing a
wide range of molecular structural features and distinguishing
among positional isomers, the technique has found increasing
interest in different fields of analysis of volatiles. Nevertheless, publications about practical applications of the GC-
FTIR technique are rather scarce as yet. In most cases, model
systems or applications in environmental analysis have been
studied (2-5). Thus, we have checked the technique for its applicability in flavour analysis, i.e. for the study of a complex
mixture of volatile compounds naturally occurring in a high dynamic range, but in rather trace concentrations. In this paper,

some results of our HRGC-FTIR investigations on tropical fruit volatiles will be presented.

Results and discussion

In our recent studies of tropical fruit volatiles we have dealt especially with the volatile constituents of fresh guava (Psidium guaiava, L.) (6,7) and cherimoya (Annona cherimolia, Mill.) (8) fruits. The scheme outlined in Fig. 1 shows the different steps of aroma investigation. The fruit pulp obtained after separation of skin and kernels was high-vacuum distilled; internal standards had been added. The aqueous distillate was then extracted using a pentane-dichloromethane (2+1) mixture (9). After careful concentration of the extract on a Vigreux column, the volatiles were preseparated by adsorption chromatography on silica gel employing a pentane-diethylether gradient (10). Three fractions of increasing polarity were eluted, each of which were analyzed in the next step by fused silica capillary gas chromatography and by coupled capillary GC-MS as well as on-line capillary GC-FTIR

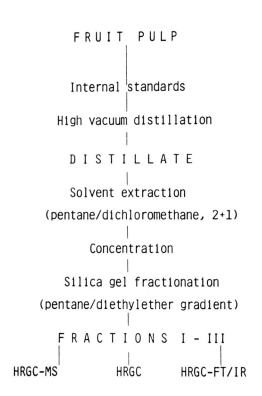

FRUIT PULP
|
Internal standards
|
High vacuum distillation
|
DISTILLATE
|
Solvent extraction
(pentane/dichloromethane, 2+1)
|
Concentration
|
Silica gel fractionation
(pentane/diethylether gradient)
|
FRACTIONS I - III
| | |
HRGC-MS   HRGC   HRGC-FT/IR

Fig. 1 Scheme of separation, concentration and analysis of tropical fruit volatiles. Details of sample preparation and analysis have already been published elsewhere (8)

spectroscopy. The residues of high-vacuum distillations were
also investigated after solvent extraction (7,8).

In the following, our interest will be focused on the HRGC-FTIR
analysis. This technique was used to analyze fractions II and
III of cherimoya and fraction II of guava fruit volatiles. Before discussing the results obtained by this technique let us
have a look at the HRGC-FTIR system used in these studies (Fig.
2).

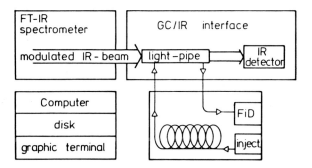

Fig. 2 Scheme of main parts of the HRGC-FTIR system (Bruker IFS
85) used. Details of analysis have already been described (11)

The sample was introduced by on-column injection technique. The
fused silica column used was a 30 m Chrompack CPWax 57 bonded
phase wide bore column. It ended directly in the heated 'light-
pipe' cell placed in the path of the modulated infrared beam
from the FTIR instrument. Parallel with the effluent from the
column make up gas was introduced into the light-pipe to maintain linear gas velocity throughout the whole GC-IR interface.
A fused silica return line guides the gas stream from the light
pipe outlet via en effluent splitter to the FID detector.

As to the results let us begin with the volatiles of cherimoya
fruit pulp. Cherimoya is a tree fruit originating from Central
America and is now cultivated there and also in different countries of South America. In Europe we get the fruit mostly from

Spain. After removal of the skin and the kernels you will find in the inside a soft pale pulp with a delicious flavour, often described as strawberry-cream or pineapple-banana. In the following we would not like to continue with sensory expressions, but rather with the chemical analysis. As already mentioned, three silica gel fractions were preseparated from which fractions II and III were studied by HRGC-FTIR. In addition, all three fractions were studied by HRGC and HRGC-MS (8). Fig. 3 shows the HRGC separation of cherimoya volatiles in fraction II. One can see that the reconstructed FTIR chromatogram (top) is in good agreement with the corresponding FID trace (bottom).

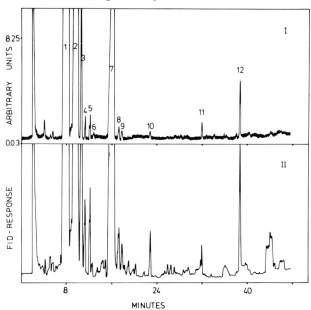

Fig. 3 Total IR (I) - and FID (II) chromatograms of silica gel fraction II of cherimoya fruit volatiles. Experimental details, see (11)

In Table 1 the quantitatively determinating part of the identified volatiles including their retention indices as well as their concentrations are outlined. The identifications were performed by comparison of GC- and IR data with those of authentic reference material. A commercially available IR vapour phase

Table 1   Cherimoya volatiles identified by HRGC-FTIR in silica
           gel fraction II and amounts (µg/kg) determined by HRGC

Peak no.[a]	Linear retention index		Compound	Amount (µg/kg)[b]
	Sample	Reference		
1	1215	1221	Butyl butanaote	490
2	1270	1267	3-Methylbutyl butanoate	300
3	1287	1287	3-Methylbutyl 3-methyl-butanoate	35
4	1303	1305	Pentyl butanoate	5
5	1324	1327	(Z)-2-Pentenyl butanoate	5
6	1336	1334	Butyl 2-butenoate	1
7	1415	1405	Hexyl butanoate	140
8	1430	1434	Hexyl 3-methylbutanoate	5
9	1440	1444	3-Methylbutyl hexanoate	2
10	1533		Unsaturated ketone	2
11	1712	1700	1,4-Dimethoxibenzene	1
12	1814	1834	Benzyl butanoate	5

[a] The peak no. correspond to the numbers in Fig. 3.

[b] Standard controlled capillary gas chromatographic determinations in fruit pulp without consideration of calibration factors, i.e. F = 1.00 for all compounds.

library (12) was used for further confirmation - where possible. As it can be seen from Table 1, the identification of all numbered peaks except one (peak no. 10) could be achieved by vapour phase infrared spectroscopy.

Which quality of IR spectra can one now expect for this substances ? To answer this question, examples of vapour phase IR spectra taken from silica gel fraction II of cherimoya volatiles (cp. Table 1) are represented in Fig. 4.

298

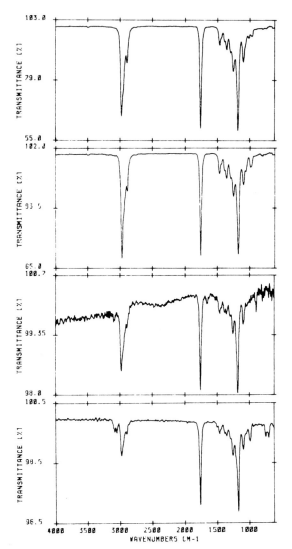

Fig. 4 Vapour phase IR spectra from HRGC-FTIR run on silica gel fraction II of cherimoya volatiles (peak no. 1,2,5,12 - from top - cp. Table 1)

All four 'on the fly' spectra show the typical absorption bands for esters with the valence band for the carbonyl function (C=O) at about 1755 cm^{-1} and of the ester band (C-O) between 1172 and 1180 cm^{-1}. The principle difference between the spectra of esters taken in the vapour state compared to other states lies in the C=O band, which is shifted towards slightly higher wavenumbers because of the lack of a 'solvent shift'. The fingerprint regions of the demonstrated spectra, i.e. the region downwards from about 1500 cm^{-1} are obviously very similar to each other as the acid function in all shown esters is the same; they are all butanoates. This demonstrates the relatively low capability of the IR spectroscopy to differentiate between homologues. Especially the first two saturated esters, butyl butanoate and 3-methyl-butyl butanoate, are difficult to discriminate by IR only, but one has to consider that using the combination of GC and IR additionally the retention data are available.

From the IR data of the next example additional absorption bands at 3082, 1650 and 895 cm^{-1} suggest an unsaturated butanoate with a Z-configuration at the double bond. The comparison with synthesized samples revealed that this compound has the structure of Z-2-pentenyl butanoate, a component of special interest for the flavour chemist as it has not been identified so far as a plant volatile. It possesses a strong 'fruity' note. It has to be mentioned that the spectrum was taken from a sample amount of somewhat less than 100 ng in the separated GC-peak.

As to the last spectrum in this series the double band at 3074 and 3040 cm^{-1} together with the absorption at 745 and 698 cm^{-1} suggests an monosubstituted aromatic ester, identified as benzyl butanoate.

As it can be seen from Fig. 3 and Table 1 nearly all the quantitatively main components in silica gel fraction II of cherimoya volatiles were detected and identified by HRGC-FTIR. Table 2 shows the main components of all three fractions. The volatile compounds which could be identified by HRGC-FTIR are marked by an asterisk (*). What we mentioned for fraction II - nearly all the main constituents could be detected by the IR technique - is also true for fraction III.

In Fig. 5 a selection of vapour phase spectra is represented from fraction III, where mostly different alcohols in relatively high concentrations could be found. From top to bottom you can see the spectra of 1-butanol, 1-penten-3-ol, 1-hexanol, li-

Table 2   Quantitative distribution of main volatile constituents of cherimoya fruit pulp

10-50	50-100	100-500	500-1000	> 1000 µg/kg[a]
α-Thujene[b]	p-Cymene[b]			
Camphene[b]				
α-Terpinene[b]				
γ-Terpinene[b]				
Limonene[b]				
3-Methylbutyl 3-methyl-butanoate *		Butyl butanoate*		
		3-Methylbutyl bu-tanoate *		
		Hexyl butanoate*		
2-Methyl-1-propanol*	2-Methyl-2-butanol*		1-Butanol*	3-Methyl-1-butanol*
2-Methyl-3-buten-2-ol	1-Pentanol*		Linalool *	1-Hexanol*
2-Pentanol	1-Penten-3-ol*			
(E)-2-Hexen-1-ol*	(E)-3-Penten-1-ol*			
2-Ethyl-1-hexanol	(Z)-2-Penten-1-ol*			
Benzylalcohol	(Z)-3-Hexen-1-ol*			
	α-Terpineol*			
Chrysanthenone				
(Z)-Linalooloxide, furanoid				

[a] Standard-controlled capillary gas chromatographic determinations in fruit pulp without considera-tion of calibration factors, i.e. F = 1.00 for all compounds.

[b] Not studied by HRGC-FT/IR.

* Identified by HRGC-FT/IR.

nalool and α-terpineol. The first and third spectrum (from top) are typical for saturated aliphatic alcohols. Characteristic for the vapour phase spectra is a sharp stretching band at about 3670 cm^{-1} indicating a monomeric free hydroxyl group in contrast to the well-known broad band for this function in the condensed phase. In all vapour phase spectra a characteristic band can be found at 1380 cm^{-1} with an increased intensity compared to the liquid phase spectra, characterizing the alcohol function.

In addition to the above mentioned absorptions the second spec-trum shows a band for unsaturated C-H at 3082 cm^{-1}. Together with the absorptions at 995 and 930 cm^{-1}, the out of plane de-formation band of a terminal double bond, and the -OH band occu-rring at 3650 cm^{-1}, a secondary 1-en-3-ol can be deduced. After comparison with reference samples 1-penten-3-ol could be confirmed. The two spectra at the bottom of the figure are repre-

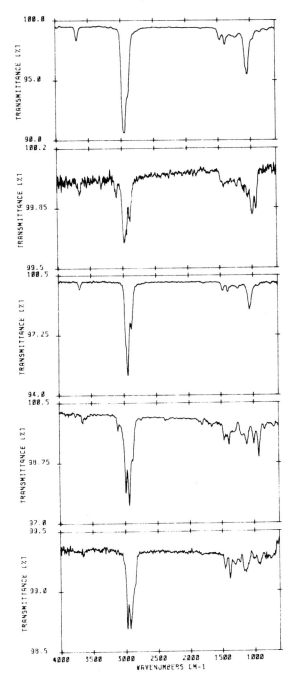

Fig. 5 Vapour phase IR spectra from HRGC-FTIR run on silica gel fraction III of cherimoya volatiles (1-butanol, 1-penten-3-ol, 1-hexanol, linalool, α-terpineol, from top)

302

sentative for tertiary terpene alcohols. Thus, the additional
absorption bands in the third spectrum exhibit the structural
elements of a terminal (1000 and 921 cm^{-1}) and a threefold substituted double bond (831 cm^{-1}), in agreement with the structure of linalool, whereas in the bottom spectrum the shoulder peak
at 3020 cm^{-1} indicates a cyclohexene ring according with the
structure of α-terpineol. When compared with authentic samples
both structures could be proved.

Let us now proceed to the results we obtained from our study of
guava. The guava is a tree or shrub grown in most tropical and
substropical countries. In Europe the fruit is mostly imported
from Brazil. It has a greenish yellow skin and is round to ovoid
in shape. Located in the center of the pale yellow or reddish
flesh that imparts a gritty texture due to the numerous stone
cells there are many
small hard seeds. The
flavour can be described
as quince-banana like.
As yet we have only stu-
died silica gel fraction
II of guava volatiles by
means of the HRGC-FTIR
method. The dominating
components of this frac-
tion are shown in Table
3. Similar to our results
obtained with cherimoya
we can say that it is possible to identify most of
the main guava volatiles
by means of HRGC-FTIR
technique. Selected exam-
ples will be provided in
the next figures. From
Table 3 it can be seen

Table 3   Main volatile constituents in silica gel fraction II of guava fruit pulp

100-500	500-1000	> 1000 µg/kg[a]
Hexyl acetate[*]		(Z)-3-Hexenyl acetate[*]
(E)-2-Hexenyl acetate		Phenylpropyl acetate[*]
Cinnamyl acetate		
Ethyl butanoate		
Ethyl 2-butenoate		
Ethyl hexanoate[*]		
Ethyl 2-hexenoate		
Methyl octanoate[*]		
Ethyl octanoate		
Methyl benzoate		
Ethyl benzoate[*]		
(Z)-Methyl cinnamate		
(E)-Methyl cinnamate		
(E)-Ethyl cinnamate		
(E,Z)-2,4-Heptadienal[*]	(E)-2-Pentenal[*]	Hexanal[*]
(E,E)-2,4-Heptadienal	(Z)-3-Hexenal[*]	(E)-2-Hexenal[*]
(E)-2-Octenal[*]		
(E)-2-Decenal		
(E,E)-2,4-Decadienal[*]		
Benzaldehyde		
4-Methylheptan-3-one		
(E)-Theaspirane		

[a] Standard-controlled capillary gas chromatographic determinations without consideration of calibration factors, i.e. F = 1.00 for all compounds.

[*] Identified by HRGC-FT/IR.

that various aldehydes are present among the guava flavour substances; as examples in Fig. 6 vapour phase IR spectra of two
aldehydes are represented. They exhibit the typical aldehyde

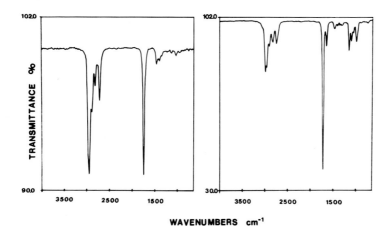

WAVENUMBERS cm⁻¹

Fig. 6 Aldehyde vapour phase IR spectra from HRGC-FTIR run on
silica gel fraction II of guava volatiles (hexanal, E-2-hexenal,
from left)

absorptions at 2805 und 2711 cm^{-1} for the saturated hexanal at
the left hand side, which is shifted to 2805 and 2721 cm^{-1} for
an α,β-conjugated example at the spectrum shown at the right
hand side. The so-called 'fingerprint' band for α,β-conjugated
aldehydes at 1141 cm^{-1} confirms this proposal. The E-configura-
tion of the double bond can be deduced from the peaks at 1632
and 982 cm^{-1}, respectively; we could identify this peak as E-2-
hexenal.

In Fig. 7 again several ester spectra are displayed, this time
from guava volatile fraction II. Here the spectra of ethyl hexanoate (top left), hexyl acetate (top right), Z-3-hexenyl acetate (bottom left), and methyl octanoate (bottom right) are
shown. For the interpretation of the saturated ester spectra
(all spectra except the spectrum at bottom left) we would like

to refer to the discussion of cherimoya esters. In the spectrum at bottom left the additional absorption at 3018 cm^{-1} and the shift of the C=O to 1764 cm^{-1} indicate an unsaturated ester.

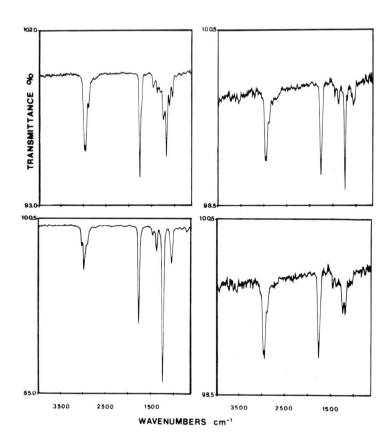

Fig. 7 Ester vapour phase IR spectra from HRGC-FTIR run on si-licagel fraction II of guava volatiles. From left to right, ethyl hexanoate, hexyl acetate (top); Z-3-hexenyl acetate, me-thyl octanoate (bottom)

Furthermore, from the band at 1047 cm^{-1} a 3-hexenyl ester can be deduced, whereas the wavenumber of the C-O absorption at re-latively high value of 1235 cm^{-1} points to an acetate. Typical for the Z-configuration of the 3-hexenyl structure is the almost complete absence of an absorption band for C=C around 1650 cm^{-1} due to the relatively high symmetry of the Z-3-hexenyl structure. In contrast to this the E-isomer shows a medium band in this region.

Conclusions

What could we demonstrate with this flavour research work ? First of all, we described the flavour composition of cherimoya fruit for the first time, and including our GC-MS analysis (6) could enlarge our present knowledge about guava volatiles. Secondly, we could show the possibilities and limits of the on-line capillary GC-FTIR technique at its present stage for its use in flavour analysis. From these results one can expect a wide range of applications in the routine and control analysis of different flavour substances. For flavour research work, i.e. the analysis in the ppb and even sub-ppb area the technique is still limited because of sensitivity problems. One should consider that e.g. the total complex of cherimoya fruit volatiles comprises more than 200 substances occurring mostly in concentrations of less than 1 ppb as we could recently demonstrate by GC-MS identification procedure (8). But one may assume that with further improvements of the chromatographic integrity of the GC-FTIR interfaces, e.g. smaller diameters and longer light-pipes for optimal adaption to the widely used WCOT capillary columns, lower detection limits should be achieved.

On the other hand, it must be pointed out that one of the advantages of this technique consists of the nondestructive character of detection. Thus, with the advent of GC-FTIR analysis of real samples comes the opportunity to supplement mass spectrometrical analyses or other techniques in on-line procedures. At present, however, a major limitation to a rapid and positive identification by GC-FTIR analysis is the lack of adequate vapour phase IR spectral data banks, but there is no doubt that they - similar to the fast development in MS analysis - will be available in the near future.

References

1. Low, M.J.D., Freeman, S.K., Anal. Chem. <u>39</u>, 194 (1967).
2. Wieboldt, R.C., Hohne, B.A., Isenhour, T.L., Appl. Spectr. <u>34</u>, 7 (1980).
3. Kuehl, D., Kemeny, G.J., Griffiths, P.R., Appl. Spectr. <u>34</u>, 222 (1980).
4. Shafer, K.H., Byorseth, A., Tabor, J., Jakobsen, R.J., HRC & CC <u>3</u>, 87 (1980).
5. Crawford, R.W., Hirschfeld, T., Sanborn, R.H., Wong, C.K., Anal. Chem. <u>54</u>, 817 (1982).
6. Idstein, H., Diss. Univers. Würzburg, in prep.
7. Idstein, H., Schreier, P., Proc. EURO FOOD CHEM II, FECS, 119, Rom, 1983.
8. Idstein, H., Herres, W., Schreier, P., J. Agric. Food Chem. in press.
9. Drawert, F., Rapp, A., Chromatographia <u>1</u>, 446 (1968).
10. Schreier, P., Drawert, F., Winkler, F., J. Agric. Food Chem. <u>27</u>, 365 (1979).
11. Herres, W., Idstein, H., Schreier, P., HRC & CC, in press.
12. Sadtler Infrared Vapor Phase Spectra, Sadtler Lab. USA.

THE APPLICATION OF HIGH AND LOW RESOLUTION MASS SPECTROMETRY IN GC/MS COUPLING FOR ANALYZING COMPLEX VOLATILE MIXTURES OF PLANT TISSUE CULTURES

Gerda Lange
Institut für Organische Chemie der Universität Würzburg
D-8700 Würzburg

Wulf Schultze
Institut für Botanik und Pharmazeutische Biologie
der Universität Würzburg
D-8700 Würzburg

Introduction

Plant tissue cultures are an important branch of research in biology and related fields. They have been used inter alia to study questions of plant physiology. Production of numerous compounds of secondary metabolism by such cultures was described (1,2). Our own investigations are dealing with the detection of volatile substances in callus cultures from essential oil plants. These cell cultures, showing neither differentiations like leaves or roots nor specialized sites of accumulation for lipophilic compounds, were able to biosynthesize very complex and heterogeneous mixtures of volatiles (3,4). Sometimes those mixtures contain more than one hundred individual substances in a wide range of concentrations and belong to very different classes of natural compounds.

The total amount of such a mixture is often very low (approximately 0.001% of wet weight) and the separation by chemical methods is not possible. Therefore, we used gas chromatography-mass spectrometry with a data system (GC/MS/DS) as a combined technique which is a powerful tool to investigate complex mix-

tures of that kind. In this paper we want to present a few selected examples of such an analysis to demonstrate especially
the possibilities of MS/DS to extract as much information as
possible.

Results

The mass spectrometry experiments were done in electron impact
(EI) and chemical ionization (CI) mode in low (LR) and high
resolution (HR). GC/MS capillary coupling in low resolution is
already well established in analytical laboratories (5,6).
High resolution mass spectrometry which gives important additional information is not yet used very often in combination
with capillary GC. However, this technique is developing since
fast scanning magnetic sector mass spectrometers connected to
data systems with a sufficient digitization rate are commercially available (5,7,8,9).

In this paper the term "high resolution" does not mean a particular numerical value but an increased resolution which allows the determination of the elemental composition of ions
from accurate mass measurement. The resolution cannot be chosen independently from other instrument operating parameters.
Equation (1) in table 1 shows the interrelationship between
digitization rate, number of data points across a mass peak,
resolution and scan rate. They all have an effect on mass spectral sensitivity, as it can be seen from table 1 (5,8,9). Too
high resolution and too fast scan rate cause a severe decrease
in sensitivity so that components of only low concentration may
not be detected in a HRMS measurement. On the other hand, mass
multiplets of only small mass differences need adequate resolution (table 2) and fast eluting capillary GC peaks demand
sufficient scan speed and short cycle time. Therefore, a compromise between the operating parameters has to be made. Resolution should be kept as low and scan rate as slow as possible.

Table 1   Relationships between operating parameters in HRMS

PARAMETER	RELATIONSHIP	REQUIREMENTS,LIMITATIONS
Resolution	Scan rate Digitization rate MS sensitivity	High resolution requires high digitization rate and slow scan rate High resolution reduces MS sensitivity
Scan rate (Cycle time)	GC-peak width MS sensitivity	Small GC-peak width requires fast scan rate,short cycle time Fast scan rate reduces MS sensitivity
Digitization rate	Resolution Scan rate Number of data points MS sensitivity	High digitization rate permits higher resolution and faster scan rate Slow digitization rate results in a better peak representation

$$\text{DIGITIZATION RATE} = \frac{\text{DATA POINTS} \cdot \text{RESOLUTION} \cdot \text{Ln10}}{\text{SCAN RATE}} \qquad (1)$$

Table 2 shows some of the basic doublets including C,H,O,S,N
which could be expected in our mixture, and the theoretical
resolution necessary for their separation. Separating a doublet
in the high mass range needs more resolving power than in the
low mass range. However, accurate mass measurement does not
need complete, theoretical multiplet separation. Suitable mass
accuracy for a molecular ion, which always is a singlet can be
attained by even a considerably lower resolution. The interference with ions of the same nominal mass but different elemental composition from overlapping GC peaks is very seldom.
However, mass measurement will be incorrect in the case of
interference of the molecular ion with an intense ^{13}C-isotope
peak of the $(M-1)^+$ fragment. Separation of the CH/^{13}C doublet
will not be possible in capillary GC/HRMS.

We used a resolution of 5000 and 6500, respectively. In general, this provides a sufficient separation of hydrocarbon and

Table 2   Basic doublets including C,H,O,S,N and the theoretical resolution necessary for their separation

ELEMENTAL DIFFERENCE	MASS DIFFERENCE (mmu)	RESOLUTION (10% VALLEY)[+] AT MASS		
		50	150	250
$S - CH_4O$	54.1	900	2700	4600
$S - {}^{13}CH_3O$	49.7	1000	3000	5000
$O - CH_4$	36.4	1500	4000	7000
$O - NH_2$	23.8	2100	6300	10500
$S - O_2$	17.7	2800	8500	14000
$N - CH_2$	12.6	4000	12000	20000
$N_2 - CO$	11.2	4500	13500	22000
$^{13}C - CH$	4.5	11000	33000	55000
$C_3 - SH_4$	3.4	15000	45000	75000

+ round values

heteroatom containing ion species to make correct statements
about the elemental composition of fragment ions, especially
in the lower mass range. There were no interferences with PFK
(perfluorokerosene) reference peaks.

The data system was used not only as a recording system but as
an aid in data processing. A powerful technique for detection
and identification of particular compounds and special classes
of compounds is "mass chromatography" (10,11,12). This technique permits the reconstruction of the ion current profiles

of selected ions as a function of spectrum index number (or time) from continuously collected data. The complete mass spectra are stored on a disk and are accessible any time after data acquisition. This technique should not be confused with "mass fragmentography" (11,12,13), which is known as SIM, SID, MIS etc. and which is often used in mixture analysis. In this case, the data of only preselected ions are recorded real time and no mass spectra are available.

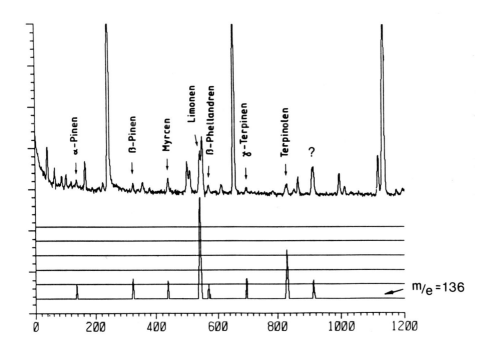

Fig. 1   Detection of monoterpene hydrocarbons in a steam distillate of a plant tissue culture by mass chromatography in LR. Monitor ion: m/e 136

Figure 1 displays the detection of monoterpene hydrocarbons in a steam distillate of a plant tissue culture. The most common ones of these terpenes are those with the elemental composition

$C_{10}H_{16}$ and the nominal molecular mass m/e 136. As far as we know, they all show a molecular ion with a definite peak in their mass spectra (14,15,16). Therefore, this ion species was used for their detection. The top trace in figure 1 shows the reconstructed ion chromatogram (RIC), which is similar to a gas chromatogram generated by a flame ionization detector. By this technique this class of substances is recognized and located very easily and quickly without the need to examine the mass spectra of all peaks in the critical part of the chromatogram. This type of chromatogram can be recorded real time as well, so one can obtain the results even before the run is finished.

A few remarks have to be made about specifity and reliability concerning ion monitoring processing in LRMS. Monitoring a nominal mass will record any ion of this mass number independently of the type of compound they are formed from or the elemental compositon, and one has to reckon with interferences. For instance, monoterpene aldehydes and alcohols eliminate water on electron impact to give $C_{10}H_{16}$-fragment ions. Many terpene esters and sesquiterpenes form this hydrocarbon fragment, too. But on a polar stationary phase, such as FFAP which was used in our experiments, those groups of compounds will hardly interfere with monoterpene hydrocarbons, as the retention time ranges are different from each other. This is not the case on apolar phases like SE 30, so that monitoring the nominal mass m/e 136 would not be possible to detect monoterpene hydrocarbons reliably. However, there are special monoterpene ethers, which form a fragment m/e 136 too, e.g. 1,8-cineole, which elute in the same retention time range as monoterpene hydrocarbons on FFAP or Carbowax phases. In our mixture we did not notice any critical interferences. But we have to emphasize that the complete spectrum must be examined in any case, when the compound has been located.

Figure 2 exhibits the detection of sesquiterpene hydrocarbons (elemental composition, $C_{15}H_{24}$) in a tissue culture distillate,

monitoring the molecular mass m/e 204. According to our knowledge, compounds of this group form a molecular ion on electron
impact, which however, might be of low intensity in some cases
(17,18). The ion current profile of mass m/e 204 shows two sesquiterpene hydrocarbons to be present, one of them only in a
small trace, which is not detected reliably in the RIC. As to
selectivity and reliability similar considerations have to be
made as above-mentioned. In the expected retention time range
we could not observe any interference with other compounds.

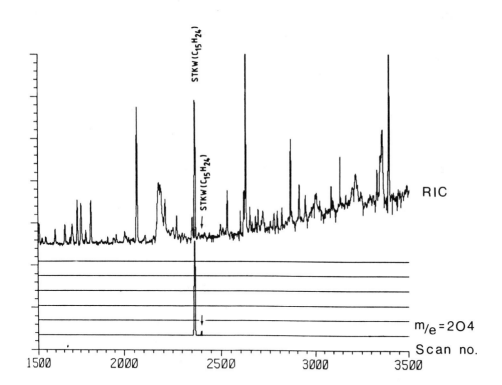

Fig. 2 Detection of sesquiterpene hydrocarbons in a steam
        distillate of a plant tissue culture by mass chromato-
        graphy in LR. Monitor ion: m/e 204

314

Ion monitoring of the molecular ion for tracing particular
classes of compounds can be applied only in some special cases.
More often, one has to use characteristic fragment ions to detect a specific group of substances as their members may cover
a wide range of molecular masses. For instance, n-alkanals
containing more than three carbon atoms undergo McLafferty rearrangement (19,20,21,22) to yield the characteristic and in
general prominent ion m/e 44 ($C_2H_4O^+$). This fragment can serve
for detection of n-alkanals by mass chromatography (Fig. 3).

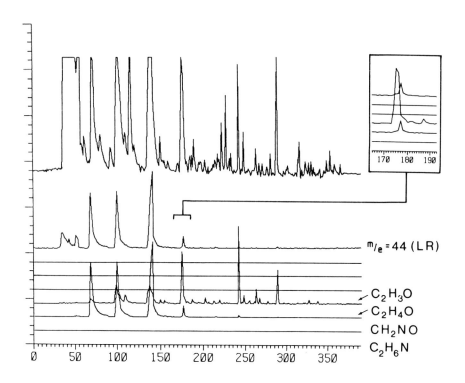

Fig. 3 Mass chromatogram of m/e 44 in low and high resolution.
Detection of n-alkanals by means of the ion species
$C_2H_4O^+$

Unfortunately, many other types of compounds form this ion of
nominal mass m/e 44 in their mass spectra too, i.e. amines,

amides and many other oxygen containing substances. To exclude
all possible nitrogen containing species of mass m/e 44, e.g.
$C_2H_6N^+$ and CH_2NO^+ (as well as any corresponding ^{13}C-species
with one H-atom less) the HRMS run was used for ion monitoring.
On the profile trace, where any of the $C_2H_4O^+$-ions are collected, there are four major peaks. These indicate the n-alde-
hydes C_6 to C_9, being confirmed by the complete mass spectra.
Under the experimental conditions used, the nitrogen containing
species $C_2H_6N^+$ and CH_2ON^+ are well separated. Their intensity
profile lines are blank. Thus, this single chromatogram demonstrates not only the presence of aldehydes but also the absence of such amines, amides and other nitrogen containing compounds which form these ions in their mass spectra.

To test the possible interference of $C_2H_4O^+$ with $^{13}CCH_3O^+$ (the
^{13}C-isotope of the acetyl ion) we simultaneously traced the
acetyl ion m/e 43 ($C_2H_3O^+$). At those positions where the aldehyde peaks ($C_2H_4O^+$) are maximizing, the intensity of the
acetyl ion is low. As the ^{13}C portion is only about 2% of the
acetyl ion itself, the contribution of ^{13}C-acetyl to the C_2H_4O-
trace is very small and nearly negligible. This can be tested
at the peak maximizing at spectrum no. 242 (in Fig. 3). This
compound displays an intense acetyl ion ($C_2H_3O^+$). The small
peak below on trace C_2H_4O mainly represents the corresponding
^{13}C-isotope species. Though these two ions cannot be separated
under the chosen experimental conditions, the error will not
be serious. Only for aldehydes in very low concentrations those
peaks have to be examined carefully.

The LR profile line which collects any ion species of the nominal mass m/e 44 shows several additional peaks at the beginning of the chromatogram. These are due to the ^{13}C-isotope
of the propyl ion $C_3H_7^+$ deriving from the solvent pentane.
They are recorded on the LR trace (m/e 44) but well separated
from the $C_2H_4O^+$ species as can be seen on the HR line.

316

The next example deals with the detection and identification of sulphur containing compounds (Fig. 4). This again demonstrates the advantage of MS ind HR mode. The ion species of mass number m/e 45, which corresponds to CHS$^+$, is a very common fragment ion in the mass spectra of many types of sulphur compounds, aliphatic as well as aromatic ones. This species is suitable for ion monitoring. With a resolution of 5000 (10% valley) used in our experiment, this species can be recorded easily without interference from any other possible ion type of mass m/e 45, except SiOH$^+$. The mass difference between SiOH$^+$ and CHS$^+$ is only 0,2 mmu, i.e. it is impossible to separate these ions in conventional mass spectrometry. Thus, HR measurements may lead to wrong results, if silica containing phases are used for GC separation.

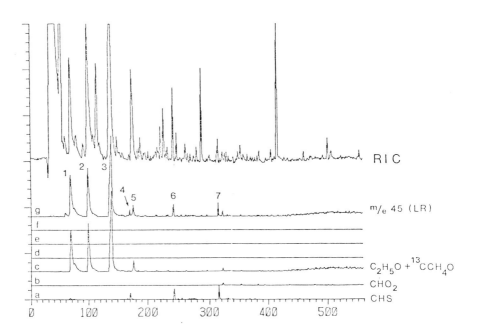

Fig. 4   Mass chromatogram of m/e 45 in low and high mass
          resolution. Differentiation between oxygen and sulphur
          containg ions

The most frequent fragments of mass m/e 45 - besides CHS^+ - are CHO_2^+, $C_2H_5O^+$ and $^{13}CCH_4O_2^+$. In Figure 4, the low resolution trace, which collects all m/e 45 ions, shows seven major peaks, only three of them (no. 4,6 and 7) are containing sulphur. For compound identification we have to examine their mass spectra and we want to select compound no. 4 for discussion. Its EI mass spectrum is depicted in Fig. 5. The HRMS data, however, gave the information that this spectrum represents two sulphur containing substances. By examining the elemental composition of the individual mass spectral peaks, it was possible to identify the two components, one being trimethylthiazol (M^+ = m/e 127, C_6H_9NS) and the other one dimethyltrisulfide (M^+ = m/e 125, $C_2H_6S_3$). The data are in agreement with the literature (23,24). These compounds were eluting at exactly the same retention time, so that clean individual spectra could not be obtained. Yet, they could be identified without GC separation mainly by the HR data.

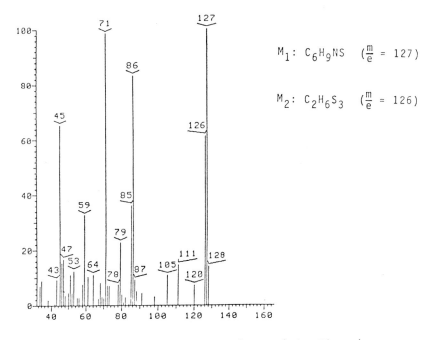

$$M_1: C_6H_9NS \quad (\frac{m}{e} = 127)$$

$$M_2: C_2H_6S_3 \quad (\frac{m}{e} = 126)$$

Fig. 5   EI mass spectrum of compound no. 4 in Fig. 4

In this case the CI run delivered incomplete information only, as the quasi M^+ (m/e 127) of the dimethyltrisulfide could not be discovered because its intensity was too low.

The resolution of 5000 was sufficient to separate CHS^+ from all other ion species of mass m/e 45. In the higher mass range, taking into account any possible elemental composition, this resolution is too low. However, the only possible elemental compositions for the ions in this spectrum are indeed those proposed by the computer. On reflection about reasonable molecules and reasonable mass differences, this can easily be confirmed. The mass measurement accuracy in this spectrum (Fig. 5) was 1 mmu on an average.

The last example demonstrates the analysis of two successively eluting GC peaks. The mass spectra plotted from the peak maxima by automatic data processing are shown in Fig. 6. Spectrum no. 2615 is easily recognized as n-octadecane by library search. But with careful examination one will discover two mass peaks, which need to be reviewed: firstly, m/e 120 cannot be rationalized for a n-alkane and secondly, m/e 111 is too high in intensity in such a hydrocarbon spectrum. These observations led to a thorough investigation of this GC peak by means of ion monitoring, CI and HRMS data. As final result we found that it consisted of three different substances (Fig. 7); two of them could be identified as n-octadecane (M^+ = m/e 254, $C_{18}H_{38}$) and ethylsalicylate (M^+ = m/e 166, $C_9H_{10}O_3$), respectively. The third one, characterized by the fragment m/e 111 is not yet identified.

The spectrum no. 2623, taken in the maximum of the second GC peak (Fig. 6) presumably represents more than one compound as well. This can be supposed from the ion sequence m/e 139, 140, 152 ending the spectrum. A detailed analysis with the aid of CI, HRMS, a slight change in GC temperature programming, and

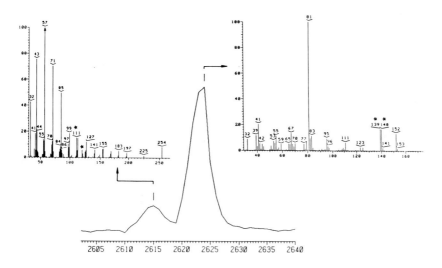

Fig. 6   Mass spectra taken from the maxima of two RIC peaks by
automatic data processing

mass chromatography led to the result   that this second GC peak
was composed of two different substances. These were identified
as 2-formyldimethylthiophene (M^+ = m/e = 140, C_7H_8OS) and 2,4-
decadienal (M^+ = m/e =152, $C_{10}H_{16}O$). Fig. 7 depicts the mass
chromatogram of those five compounds discussed above, which are
traced by their molecular ions, and the fragment m/e 111, respectively.

Conclusions

As seen by the examples described in this paper, it is obvious
that mass chromatography in low and especially high mass resolution is a powerful technique in complex mixture analysis. HRMS
gives valuable additional information regarding elemental composition, which is important for compound identification. But
one has to keep strictly in mind the limitations of this technique combined with capillary GC.

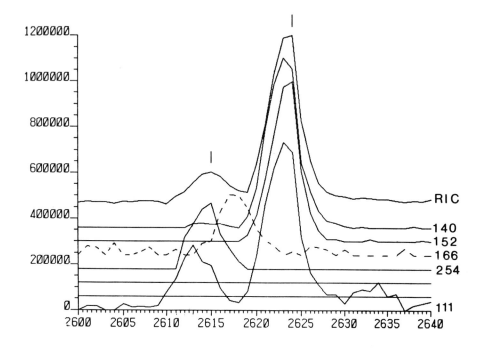

Fig. 7   Mass chromatogram of five compounds hidden under two
          GC peaks (for details, see text)

Experimental

MS conditions for the examples in Fig. 1-5: Mass spectrometer,
Varian MAT 212; data system SS 200; EI/LR and CI/LR: resolution
1000, scan rate 1.8 sec/dec; digitization rate, 9.76 KHz; CI
reactant gas, isobutane; EI/HR: resolution 5000, scan rate 6.0
sec/dec; digitization rate, 19.53 KHz.

MS conditions for the examples in Fig. 6 and 7: Mass spectrometer Finnigan MAT 8200; data system SS 300; EI/LR and CI/LR: resolution 1000; scan rate 0.8 sec/dec; digitization rate 50 KHz;
CI reactant gas, isobutane; EI/HR: resolution 6500; scan rate
2 sec/dec; digitization rate 100 Khz.

GC conditions:

Gas chromatograph: Varian 3700.

Column: 50 m glass capillary, FFAP

Injection port: Temperature 210°. Injection: Split 1:20

Column temperature: 65° (8' isotherm), increasing up to
                    220° (heating rate: 3°/min)

Carrier gas: He, flow: 2 ml/min

Transfer line: Type: open coupling with a fused silica capil-
               lary (220°)

We would like to express our thanks to Finnigan MAT, Bremen for performing measurements, which were done with regard to mass spectrometer tests. Data processing and interpretation was done in our laboratory.

References

1. Butcher, D.N.: Secondary Products in Tissue Cultures, in
    Applied and Fundamental Aspects of Plant Cell, Tissue and
    Organ Culture, J. Reinert and Y.P.S. Bajaj (Eds.), Springer
    Verlag, Berlin, Heidelberg, New York 1977, pp. 668-717

2. Staba, E.J. (Ed.): Plant Tissue Culture as a Source of Bio-
    chemicals, CRC Press Inc., Boca Raton, Florida 1980

3. Schultze, W.: Thesis, Würzburg 1982

4. Schultze, W., Koch, I., Czygan, F.-C.: Dtsch. Apoth. Ztg.,
    in press

5. McFadden, W.: Techniques of Combined Gas Chromatography/Mass
    Mass Spectrometry: Applications in Organic Analysis, Wiley-
    Interscience Publication, J. Wiley and Sons, New York,
    London, Sydney, Toronto 1973

6. Gudzinowicz, B.J., Gudzinowicz, M.J., Martin, H.F.: Funda-
    mentals of Integrated GC-MS, Part II and III, Dekker Inc.,
    New York 1976 and 1977

7. Rapp, U., Schröder, U., Meier, S., Elmenhorst, H.: Chroma-
    tographia 8, 474-478 (1975)

8. Kimble, B.J.: Introduction to Gas Chromatography/High Re-
    solution Mass Spectrometry in High Performance Mass Spec-
    trometry, M.L. Gross (Ed.), ACS Symposium Series 70, Am.
    Chem. Soc., Washington D.C. 1978, pp. 120-149

9. Chapman, J.R.: Computers in Mass Spectrometry, Academic Press, London. New York. San Francisco 1978

10. Hites, R.A., Biemann, K.: Anal.Chem. 42, 855-860 (1970)

11. Fenselau, C.: Anal.Chem. 49, 563A-570A (1977)

12. Budde, W.L., Eichelberger, J.W.: J.Chromatogr. 134, 147-158 (1977)

13. Falkner, F.C., Sweetman, B.J., Watson, J.T.: Appl. Spectrosc. Rev. 10, 51-116 (1975)

14. Ryhage, R., Sydow, E.: Act.Chem.Scand. 17, 2025-2035 (1963)

15. Thomas, A.F., Wilhalm, B.: Helv.Chim. Acta 47, 475-488 (1964)

16. Bünau, G., Schade, G., Gollnick, K.: Z.Anal.Chem. 244, 7-17 (1969)

17. Moshonas, M.G., Lund, E.D.: Flavour Ind. 1, 375-378 (1970)

18. Stenhagen, E., Abrahamsson, S., McLafferty, F.W.: Atlas of Mass Spectral Data, Vol.1, J.Wiley and Sons, New York. London. Sydney. Toronto 1974

19. Gilpin, J.A., McLafferty, F.W.: Anal.Chem. 29, 990-994 (1957)

20. Liedtke, J.R., Djerassi, C.: J.Am.Chem.Soc. 91, 6814-6821 (1969)

21. Harrison, A.G.: Org.Mass Spectrom. 3, 549-555 (1970)

22. Meyerson, S., Fenselau, C., Young, J.L., Landis, W.R., Selke, E., Leitch, L.C.: Org. Mass Spectrom. 3, 689-707 (1970)

23. Tabacchi, R.: Helv.Chim.Acta 57, 324-336 (1974)

24. Schreyer, J.: J.Agr.Food Chem. 24, 336-341 (1976)

FORMATION AND ANALYSIS OF OPTICALLY ACTIVE AROMA COMPOUNDS

Roland Tressl and Karl-Heinz Engel
Forschungsinstitut für Chemisch-technische Analyse
Technische Universität Berlin
Seestrasse 13, D-1000 Berlin 65

Introduction

Gas chromatographic resolution of chiral components can be achieved by two methods:

1. separation of enantiomers on an optically active stationary phase

2. formation of diastereoisomeric derivatives and analysis on a non-chiral phase.

The development of thermally stable chiral phases and suitable derivatization procedures made possible the separation of a broad spectrum of optically active compounds (1). Formation of diastereoisomeric derivatives for separation on non-chiral phases has been carried out with a large number of reagents (2). In this paper we want to demonstrate the application of R(+)-α-methoxy-α-trifluoromethylphenylacetic acid (R(+)-MTPA) as reagent for derivatization of chiral components and following capillary gas chromatographic separation. R(+)-MTPA-chloride, known in literature as "Mosher's reagent" (3), was developped as reagent for formation of diastereoisomeric esters and amides followed by NMR-analysis. The development of capillary gas chromatography and the increased power of resolution has now made possible the gas chromatographic separation of diastereoisomeric R(+)-MTPA-derivatives of different classes of chiral hydroxy-compounds. As shown in Figure 1 we investigated derivatives of secondary alcohols, esters of 2-, 3-, and 4-hydroxyacids and the corresponding γ-lactones.

The advantage of gas chromatographic investigation of formed
derivatives is the possibility of micro-scale procedure and
analysis at trace level, when neither measuring of the optical
rotation nor NMR-analysis are possible.
This method - experimental data will be published in detail -
was applied to control the formation of optically active compounds during microbiological processes and to determine the
enantiomeric composition of chiral aroma constituents in
natural systems at trace level.

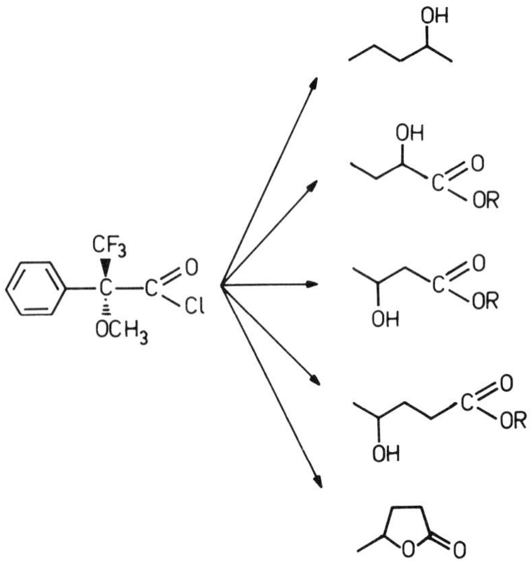

Figure 1:   Chiral hydroxy-compounds derivatized
            with R(+)-MTPA-Cl

Analysis of secondary alcohols

Gas chromatographic separation of secondary alcohols has been
achieved by investigation of different diastereoisomeric derivatives (4,5). Attygalle et al. (6) recently compared the
efficiency of several derivatization procedures for separation
of octanol-3. König et al. (7) resolved chiral alcohols after
formation of isopropylurethanes on a chiral phase.

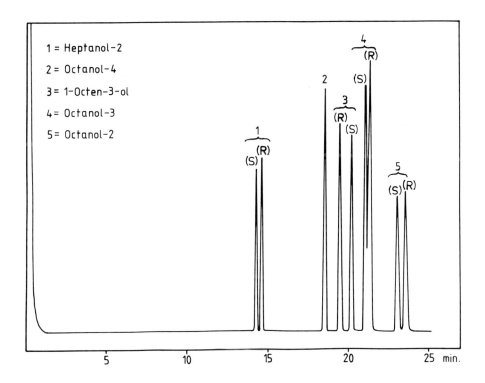

Figure 2:   Capillary GC-separation (OV-101, 50m/0.32mm i.d.,
            170°C) of R(+)-MTPA-derivatives of some secondary
            alcohols

Figure 2 presents the capillary GC-separation of diastereoisomeric R(+)-MTPA-esters of some secondary alcohols. It can be
seen, that the bulk dissymmetry at the asymmetric C-atom
strongly effects the resolution. Octanol-2 is resolved better

than heptanol-2, whereas the difference in only one C-atom between the two branches in octanol-4 is not sufficient to lead to a separation. 1-Octen-3-ol, possessing the more fixed double bond, is separated best.

This method to determine the enantiomeric composition of chiral alcohols was applied to microbiological processes, which so far were controlled by measuring of the optical rotation or NMR-analysis.
The stereospecific reduction of unsymmetrical ketones by actively fermenting yeasts is well known (8).
Reduction of methylketones by Saccharomyces cerevisiae leads to the corresponding S(+)-alkan-2-ols (9). Another method to obtain optically pure alcohols is the enzymic stereospecific hydrolysis of racemic acetates by microorganisms (10).

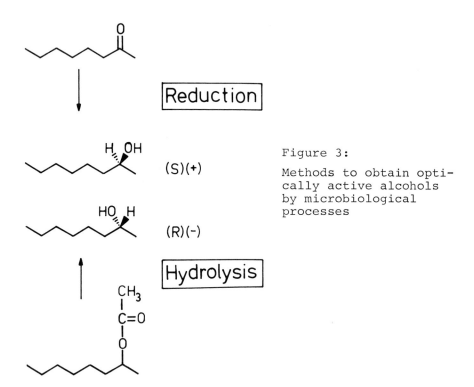

Figure 3:

Methods to obtain optically active alcohols by microbiological processes

We could show that in contrary to the reduction of ketones by
yeast, which mainly yields the corresponding S(+)-alcohol,
stereospecific enzymic resolution of racemic acetates leads to
the R(-)-alkan-2-ols (Figure 3). As example Figure 4 presents
the gas chromatographic resolution of racemic octanol-2 and
the enantiomeric composition, which we obtained by hydrolysis
of R,S-2-octylacetate by Candida utilis. The R(-)-enantiomer
is formed with an optical purity of 93 %. Similar results were
obtained by hydrolysis with Brevibacterium ammoniagenes.

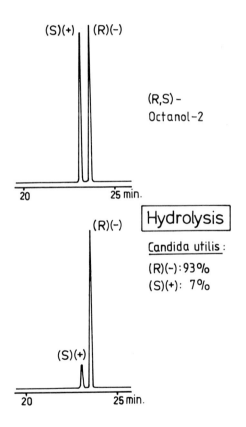

Figure 4: Stereospecific hydrolysis of racemic 2-octyl-
acetate by Candida utilis

328

Odd-numbered secondary alcohols (pentanol-2, heptanol-2, nona-
nol-2) are contained as aroma components in purple and yellow
passion fruits and we could demonstrate that the corresponding
esters are typical constituents only of the purple variety (11).

Table I: Esters identified in purple and yellow passion fruits

	(µg/kg)	
	purple	yellow
ethyl butanoate	9.000	5.300
hexyl hexanoate	4.200	4.500
octyl butanoate	600	50
2-pentyl octanoate	90	–
2-heptyl butanoate	1.000	–
2-nonyl butanoate	50	–
(Z)-3-octenyl butanoate	50	–
(Z)-3-octenyl octanoate	80	–
(Z)-3-decenyl hexanoate	50	–
citronellyl hexanoate	100	10
geranyl butanoate	100	–
geranyl hexanoate	200	–

Figure 5 presents the GC-separation of racemic pentanol-2 and
heptanol-2 and shows the enantiomeric composition of these
alcohols, isolated by preparative GC from yellow passion fruit.
Both mainly consist of the S(+)-enantiomer, the optical purity
of heptanol-2 (86%) is higher than the purity of pentanol-2
(67%). If postulating a biosynthesis of these components by reduction of the corresponding methylketones (Figure 6), the
differences in the enantiomeric composition may be explained
by the different bulk dissymmetry in both components, which
influences the steric course of ketone reduction (9).

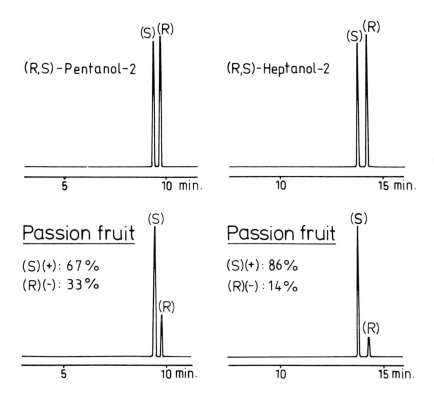

Figure 5: Enantiomeric composition of pentanol-2 and
heptanol-2 isolated from yellow passion fruit

The common principle that alcohol enantiomers obtained by
NADH-dependent ketone reduction and by asymmetric ester hydrolysis are opposite could also be confirmed for 1-octen-3-ol.
Mushrooms contain the levorotatory R(-)-1-octen-3-ol as
character impact compound. Our experiments (12) showed that
this alcohol together with some other C_8-components is formed
by lipidoxidation of linoleic acid in mushroom homogenates
(Table II). The last step of its biosynthesis is a stereospecific enzymic reduction of the corresponding keto-compound

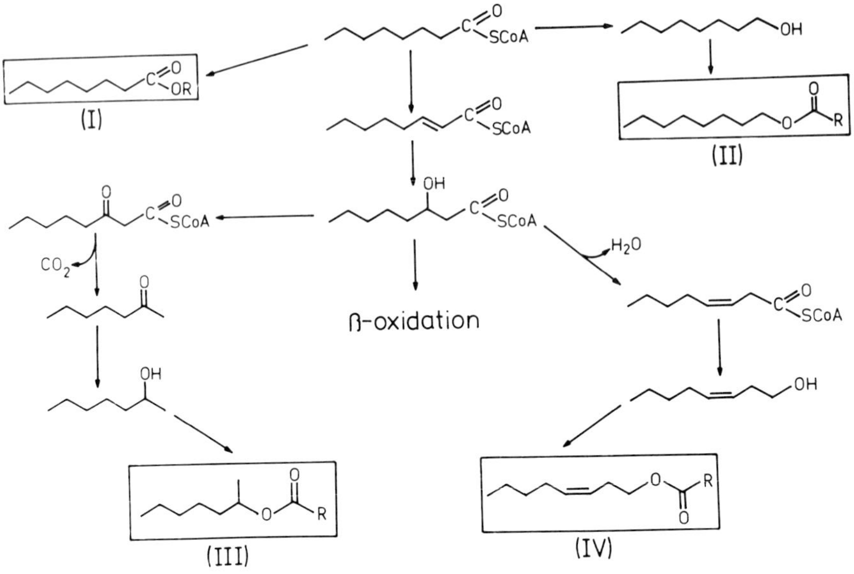

(I)

(II)

β-oxidation

CO_2

H_2O

(III)

(IV)

Figure 6: Pathway which may explain the biosynthesis of chiral alcohols and their esters in passion fruit

Table II: Formation of C_8-components in mushrooms

COMPONENTS	I	LINOLEIC ACID II	LINOLENIC ACID III
HEXANAL	0,07	0,17	0,03
3-OCTANONE	2,7	2,4	1,8
3-OCTANOL	0,45	0,46	0,39
1-OCTEN-3-ONE	0,22	0,39	0,36
1-OCTEN-3-OL	12,0	25,6	7,8
2-OCTENAL	+	+	-
2-OCTEN-1-OL	1,52	5,3	1,6
1,5-OCTADIEN-3-ONE	+	+	0,10
1,5-OCTADIEN-3-OL	+	+	10,4
2,5-OCTADIENAL	-	-	+
2,5-OCTADIEN-1-OL	+	+	5,4

1-octen-3-one to R(-)-1-octen-3-ol. In contrary to that
asymmetric hydrolysis of racemic 1-octen-3-yl acetate by Candida utilis yielded the opposite S(+)-enantiomer (Figure 7).

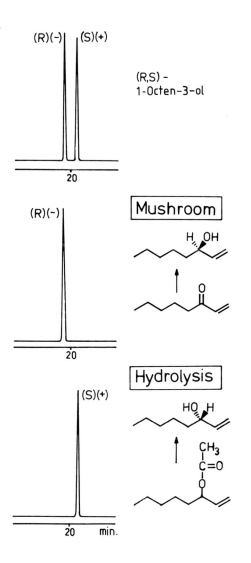

Figure 7:

Formation of opposite
enantiomers of 1-octen-
3-ol during biosynthesis
in mushrooms and enzymic
hydrolysis of 1-octen-3-
yl acetate by Candida
utilis

Analysis of esters of 2- and 3-hydroxyacids

Because the asymmetric center in 2-hydroxyacids is next to the carboxylic group, a separation of R(+)-MTPA-derivatives is possible even on a packed column (13). As example we investigated the microbiological reduction of ethyl 2-oxobutanoate by baker's yeast and could show that the obtained 2-hydroxyacidester mainly consists of one enantiomer (82 % optical purity).

3-Hydroxyacids were separated on optically active stationary phases after formation of carbamate/amide derivatives (14) or 3-pentylesters (15). Resolution of diastereoisomers was achieved by derivatization with S(-)-phenylethylisocyanate or (-) camphanic acid chloride (16). Figure 8 presents the capillary GC-separation of R(+)-MTPA-derivatives of ethyl 3-hydroxybutanoate. The order of elution was determined by comparison with the enantiomeric mixture obtained by reduction of ethyl acetoacetate with baker's yeast. It is well known (17) that this reduction leads to S(+)-ethyl 3-hydroxybutanoate with an optical purity of more than 85 %. The opposite R(-)-enantiomer can be obtained by using of a dihydroxyacetone reductase from Mucor javanicus as shown by Hochuli et al. (18), who also investigated R(+)-MTPA-derivatives.

During our studies on the aroma composition of tropical fruits we identified ethyl 3-hydroxybutanoate as constituent in mango and passion fruit. Its isolation from passion fruit by preparative GC followed by derivatization with R(+)-MTPA-Cl revealed that the enantiomeric composition of 82 % S(+) and 18 % R(-) is comparable to that obtained by yeast reduction from acetoacetate (Figure 8).

Lemieux and Giguere (19) demonstrated that reduction of the homologous 3-ketohexanoic acid by fermenting yeast leads to the opposite R(-)-3-hydroxyhexanoic acid. We could confirm these results by yeast reduction of the corresponding ethylester. GC-separation of R(+)-MTPA-derivatives showed an enantiomeric excess of 84 % R(-)-ethyl-3-hydroxyhexanoate (Fig.9).

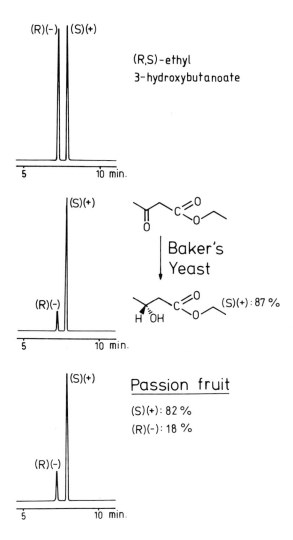

Figure 8: Enantiomeric composition of ethyl 3-hydroxy-
butanoate obtained by yeast reduction of ethyl
acetoacetate and isolated from yellow passion
fruit (GC-conditions: DB 210 (J&W) 30m/0.33mm
i.d., 180°C isothermal)

334

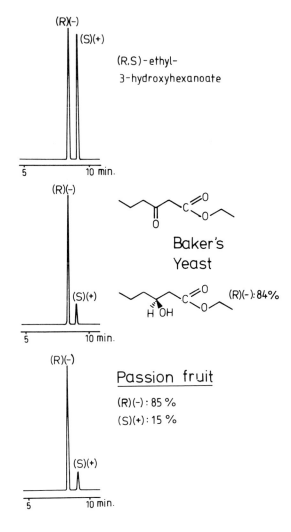

Figure 9:   Enantiomeric composition of ethyl 3-hydroxy-
            hexanoate obtained by yeast reduction of ethyl
            3-ketohexanoate and isolated from yellow
            passion fruit. (GC-conditions: DB 210 (J&W)
            30m/0.33mm i.d., 185°C isothermal)

The formation of opposite enantiomers from 3-ketobutanoic and
-hexanoic acid may be explained by the hypothesis of Benner
(20), that a difference in the reactivity of the two carbonyl-
compounds brings about the transfer of either the pro-S or the
pro-R hydrogen from syn- or anti-NADH.
Surprisingly ethyl 3-hydroxyhexanoate isolated from yellow
passion fruit also mainly consisted of the R(-)-enantiomer and
was opposite to the corresponding ethyl 3-hydroxybutanoate
(Figure 9). These findings suggest that 3-hydroxyacidesters in
passion fruit may be biosynthesized by reduction of the corresponding ketoacids by an enzyme system comparable to that of
yeast. The alternative possibility is a stereospecific hydra-
tion of Δ 2.3-trans-enoyl-CoA to S(+)-3-hydroxybutanoate during fatty acid ß-oxidation and reduction of 3-ketohexanoyl-S-
ACP to R(-)-3-hydroxyhexanoate during fatty acid biosynthesis.

Analysis of esters of 4-hydroxyacids and γ-lactones

As far as the authors know gas chromatographic resolution of
enantiomers of 4-hydroxyacids and unsubstituted γ-lactones
has not been accomplished so far. Our first experiments to separate 4-hydroxyacids after conversion to diastereoisomeric
esters by esterification with optically active alcohols failed.
The distance between the two optical centers in the formed
diastereoisomers is too far. We therefore derivatized the
hydroxyl-group at the asymmetric C_4-atom with R(+)-MTPA-Cl.
Derivatives of 4-hydroxyacid-ethylesters showed the best separation. As demonstrated in Figure 10 γ-lactones were opened
by interesterification with sodiumethylate, and the formed
4-hydroxyacidethylesters were converted to R(+)-MTPA-deriva-
tives. Figure 11 presents the capillary GC-separation of
γ-valerolactone, γ-hexalactone, and γ-octalactone. Although
no base-line separation was reached, the new method could by
applied to control the microbiological production of optically
pure γ-lactones and 4-hydroxyacid esters.

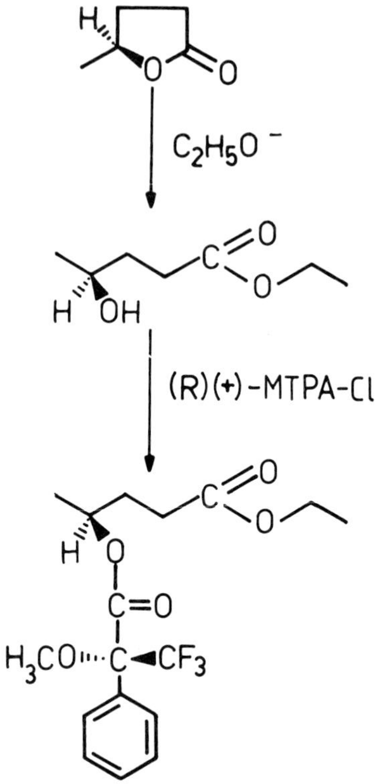

Figure 10: Conversion of γ -lactones to R(+)-MTPA-
derivatives of the corresponding 4-hydroxy-
acid ethylesters

The reduction of 4- and 5-ketoacids to optically acitve γ-
and δ -lactones by Saccharomyces cerevisiae has been described
by Muys et al. (21). We investigated the reduction of ethyl 4-
ketopentanoate. As shown in Figure 12 chemical reduction by
$NaBH_4$ yielded racemic ethyl 4-hydroxypentanoate, which could
be separated by capillary GC after reaction with R(+)-MTPA-Cl.

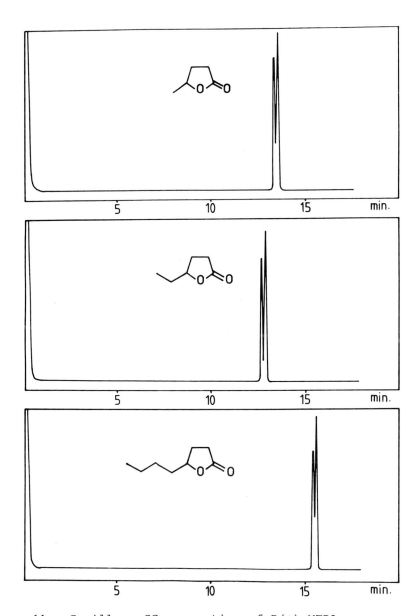

Figure 11: Capillary GC-separation of R(+)-MTPA-
derivatives of chiral γ -lactones.
GC-conditions: DB 210 (J&W), 30m/0.33mm i.d.
γ -valerolactone 175°C, γ-hexalactone 180°C,
γ -octalactone 185°C isothermal.

338

In contrary to that reduction of the keto-compound by baker's yeast led to a optically pure ethyl 4-hydroxypentanoate.

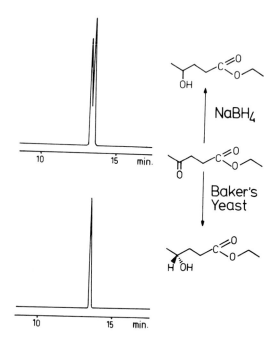

Figure 12: Formation of optically pure ethyl 4-hydroxy-pentanoate by reduction of the corresponding ketoester with baker's yeast

Similar results were obtained with 4-ketohexanoic acid. As shown in Figure 13 reduction with NaBH$_4$ yielded racemic γ-hexalactone, whereas derivatization with R(+)-MTPA-Cl and capillary GC-separation of the product obtained by reduction with baker's yeast revealed the presence of an optically pure γ-hexalactone.

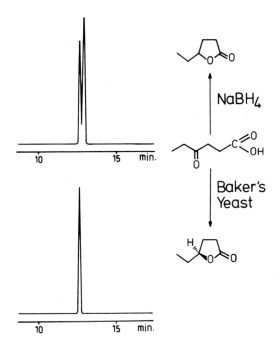

Figure 13: Formation of optically pure γ-hexalactone by
reduction of the corresponding 4-ketoacid with
baker's yeast

Analysis and separation of δ-lactones

The procedure applied to separate the optical isomers of
γ-lactones - ring-opening to the corresponding 4-hydroxyacid-
ethylester and following derivatization with R(+)-MTPA-Cl -
was not succesful for the separation of δ-lactones. If
looking at the degree of separation of R(+)-MTPA-diastereo-

isomeric derivatives of hydroxyacids, starting from 2-hydroxy-
acids, which can be resolved on packed columns, to 4-hydroxy-
acids, which require an effective capillary column, it becomes
obvious, that the separation is strongly effected by the
distance between the carboxylic group and the hydroxyl group
at the asymmetric C-atom. In 5-hydroxyacids this distance is
too far, and no separation could be accomplished. Therefore
we used a method developped by Saucy et al. (22) to determine
the enantiomeric purities of chiral δ-lactones. As shown in
Figure 14 δ-lactones were converted to their ortho esters
with D(-)-2.3-butanediol, followed by capillary GC-investi-
gation of the formed diastereoisomers. We modified the method

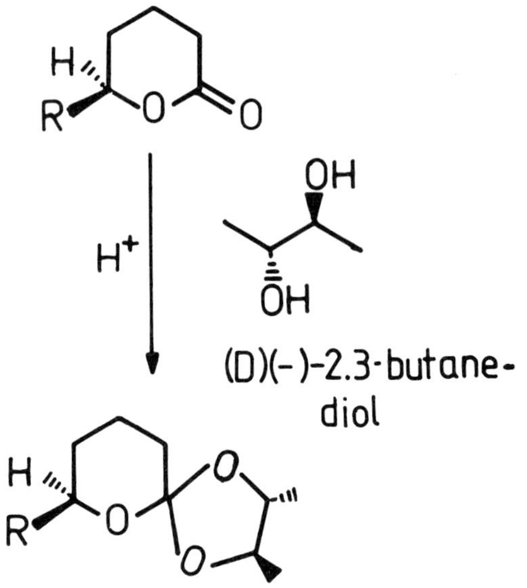

Figure 14: Conversion of δ-lactones to diastereoisomeric ortho esters by reaction with
D(-)-2.3-butanediol

to a micro-scale procedure and determined the enantiomeric
composition of δ-lactones in natural systems. We isolated
δ-octa- and δ-decalactone from a coconut cream. As shown in
Figure 14 GC-separation of the ortho esters revealed an enantiomeric purity of 92 %, and 84 %, respectively, of the two
lactones.

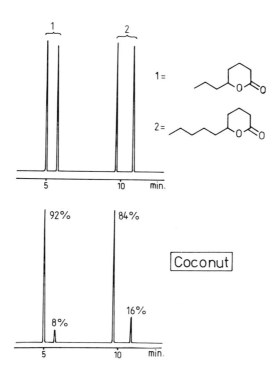

Figure 15: Enantiomeric composition of δ-octa- and
            δ-decalactone isolated from a coconut cream.
            GC-conditions: CP Wax 57, 50m/0.32mm i.d.,
            130-170°C, 2°/min.

References

1. König, W.A.: Journal of HRC & CC 5, 588-595 (1982)

2. Gil-Av, E., Nurok, D.: Adv. Chromatogr. 10, 99-172 (1974)

3. Dale, J.A., Dull, D.L., Mosher, H.S.: J. Org. Chem. 34 (9), 2543-2549 (1974)

4. Rose, H.C., Stern, R.L., Karger, B.L.: Anal. Chem. 38, 469-472 (1966)

5. Pereira, W., Bacon, V.A., Patton, W., Halpern, B.: Anal. Letters 3, 23-28 (1970)

6. Attygalle, A.B., Morgan, E.D., Evershed, R.P., Rowland, S.J.: J. Chromatogr. 260, 411-417 (1983)

7. König, W.A., Francke, W., Benecke, I.: J. Chromatogr. 239, 227-231 (1982)

8. Neuberg, C.: Adv. Carbohydrate Res.: 10, 75-117 (1949)

9. Mac Leod, R., Prosser, H., Fikentscher, L., Lanyi, J., Mosher, H.S.: Biochemistry 3, 838-846 (1964)

10. Ohta, H., Tetsukawa, H.: Agric. Biol. Chem. 44, 863-867 (1980)

11. Engel, K.-H, Tressl, R.: Chem. Mikrobiol. Technol. Lebensm. 8, 33-39 (1983)

12. Tressl, R., Bahri, D., Engel, K.-H.: J. Agric. Food Chem. 30, 89-93 (1982)

13. Kajiwara, T., Hatanaka, A., Naoshima, Y.: Agric. Biol. Chem. 44, 437-438 (1980)

14. König, H.A., Benecke, J., Lucht, N., Schmidt, E., Schulze, J., Sievers, S.: Vth Intern. Symposion on capillary chromatography, Riva del Garda, 609-618 (1983)

15. Koppenhoefer, B., Allmendinger, H., Nicholson, G., Bayer, E.: J. Chromatogr. 260, 63-73 (1983)

16. Wipf, B., Kupfer, E., Bertazzi, R., Leuenberger, H.G.W.: Helv. Chim. Acta 60 (2), 485-488 (1983)

17. Friedmann, E.: Naturwissenschaften, 400 (1931)

18. Hochuli, E., Taylor, K.W., Dutler, H.: Eur. J. Biochem. 75, 433-439 (1977)

19. Lemieux, R.U., Giguere, J.: Canad. J. Biochem. 29, 678-690 (1951)

20. Benner, E.: Experientia 38, 633-637 (1982)

21. Muys, G.T., v.d. Ven, B., de Jonge, A.P.: Nature, 995-996 (1962)

22. Saucy, G., Borer, R., Trullinger, D.P., Jones, J.B., Lock, K.P.: J. Org. Chem. 42, 3206-3207 (1977)

STEREOISOMERS OF FRUIT FLAVOUR SUBSTANCES - SOME ASPECTS OF SYNTHESIS AND ANALYSIS

A. Mosandl   and G. Heusinger
Institut für Pharmazie und Lebensmittelchemie der Universität
Würzburg, Am Hubland, D-8700 Würzburg

Introduction

In spite of considerable efforts devoted to the study of aroma
chemistry a lot of fundamental questions concerning odour qualities and odour thresholds are unresolved. As to the visual
sensations, one can see that it is possible to predict with a
high degree of accuracy the likely colour of a chemical compound
from its structural features. In the field of aroma components
such a type of prediction is nearly impossible and syntheses of
new structures with particular required flavour notes are rather difficult (1).
Nevertheless, the currently most convincing ideas on structure-
activity relationship are based on some principal conditions,
e.g. suitable physical and chemical properties like volatility
and solubility in aqueous and lipophilic media, respectively,
and one (or more) functional groups. The overall shape and size
of molecule and its detailed stereochemistry are discussed as
the most important factors. Influence of stereochemistry to
odour quality may correlate with geometric, conformational or
optical isomerism. Our own contribution to structure-function
relation will be demonstrated by the following examples.

Geometric isomers

It is well-known that the aroma note of anise is characterized
by the E-isomer of anethol. Good quality anise oils contain

less than one percent of Z-anethol, which is suggested to be hazardous to human health. Recently we were successful in LC-separating the Z-isomer from anise oil, irradiated by a mercury lamp (240-580 nm). Structure elucidation was carried out by comparison of propenyl side chain of both geometric isomers in ^{1}H-NMR spectra (Fig. 1a,1b).

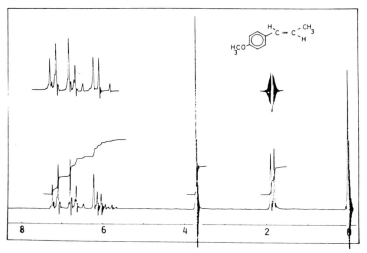

Fig.1a 60 MHz ^{1}H-NMR spectra of E-isomer of anethol

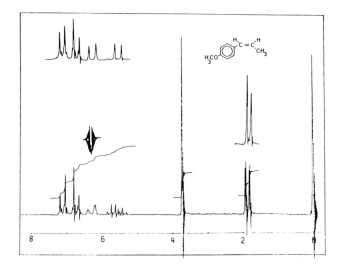

Fig. 1b 60 MHz ^{1}H-NMR spectra of Z-isomer of anethol

The methyl group of this substituent appears as a doublet in case of E-isomer and a double doublet in Z-isomer at 1,85 ppm, caused by geminal coupling with H-2 and long range coupling with H-3. Irradiation to methyl group resonance causes AB-spinning systems for the vinylic protons. Their coupling constants of J = 15,75 Hz (11,2 Hz) correspond to E- and Z-configuration, respectively. Z-isomer has a woody, minty, pungent odour, and is quite different from that of the E-isomer (odour), which exclusively correlates with the typical "anise"-note.

Optical Isomers

Three examples from the literature may demonstrate the importance of chirality to odour quality. One of the best known examples is menthol. Eight stereoisomers of menthol are known, but only the laevorotatory isomer with (1R, 3R, 4S)-configuration is the unique with the well-known fresh, minty odour.
In case of carvone two distinct odour qualities are found, both occurring in nature. Very carefully purified (-)carvone was found to have spearmint aroma, whereas ultra-pure (+)carvone is the impact substance of caraway.
Finally, only the dextrorotatory nootkatone has the classic aroma of grapefruit with its fresh, green, sour and fruity character.

Strawberry Aldehydes

3-Arylglycidic esters, the so-called "strawberry aldehydes", are known as dominant components of artificial strawberry essences (2,3,4). The application of such substances is prohibited by German food law since 1970, nevertheless, our interest is directed to these substances, because their odour quality may be affected by geometric isomerism as well as by chirality (Fig. 2) (5).

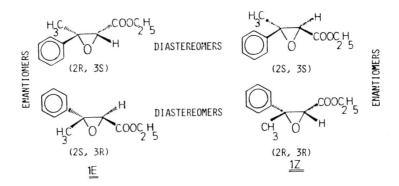

Fig.2 Stereoisomers of ethyl 3-methyl-3-phenylglycidate

In ^{1}H-NMR spectra of commercially available "strawberry aldehyde" the substituents of C-2 give rise to distinct signals,
two singulets for H-2 and two quartets and triplets, respectively, for the ethyl ester group. Hence it follows that the
usual preparation of ethyl 3-methyl-3-phenylglycidate (1) is
present as a mixture of E-/Z-isomers.
We succeeded in separating these geometric isomers by liquid
chromatography (6) on silica with petrolether/ether eluents,
and structure assignments were derived from ^{1}H-NMR spectra
(Fig. 3a,b).

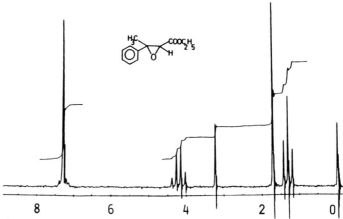

Fig.3a ^{1}H-NMR spectra of geometric isomers of ethyl 3-methyl-
3-phenylglycidate 1E

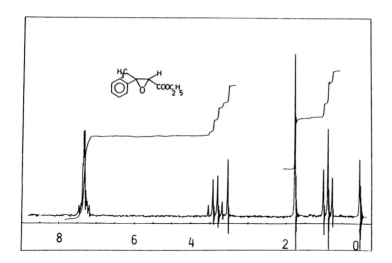

Fig. 3b ^{1}H-NMR spectra of geometric isomers of ethyl 3-methyl-
3-phenylglycidate 1Z

The different chemical shifts are caused by the shielding effect
of the phenyl ring in C-3 position. Therefore 1H-singulet of
C-2 resonates at higher field in E-isomer (1E), while in the
other case the ethyl ester group of Z-isomer (1Z) is shifted
up to a higher field.

Sensory differences of the isomers are easily to recognize.
The strawberry-like odour is caused by Z-isomer exclusively,
while the odour of the E -isomer is fruity, but not specific.
Some further glycidic esters were synthetized and separated,
and sensory qualities are listed in table 1. Between E- and
Z-isomers of glycidic esters principal sensory differences
exist.

	Z-isomer	E-isomer
	strawberry, intensive	unspecific strong
	fruity, resembling to strawberry	unspecific intensive
	strawberry, faint, unstable	sweet, must-like intensive
	faint, fruity	faint, sweet, unspecific

Table 1 Sensory qualities of some geometric isomers of "strawberry aldehydes"

For further investigations with regard to chirality and odour
quality Z-ethyl 3-methyl-3-phenylglycidate with its distinct
strawberry-odour was tested as a model substance. New procedures for optical resolution were applied, e.g. chromatography
on polymers of chiral acrylamides (7), on peracylated poly-
saccharides like cellulose (8,9), starch and chitin (5) - but
all these efforts were unsuccessful. The enantiomers of "strawberry aldehyde" finally were achieved indirectly via diastereomeric (-)menthyl glycidic esters ($\underline{2}$), which were hydrolyzed
after LC-separation (SiO$_2$; hexane/ether) and the glycidic salts
reesterificated (ethyljodide/DMF) to odour active ethyl glycidates.
Purity control results from capillary GC of diastereomeric (-)
menthyl esters with baseline resolution for Z-isomers (Fig. 4).

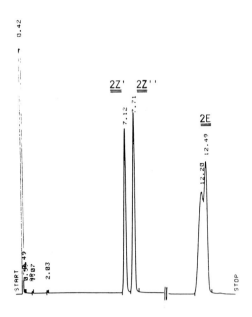

Fig.4 Diastereomeric (-)menthyl 3-methyl-3-phenylglycidates
       (Z-isomers 2Z' and 2Z'' resolved; E-isomers partially
       separated)

On the other hand enantiomeric purity of the ethyl glycidates
was determined by ^{1}H-NMR-spectroscopy with chiral shift reagents, e.g. Eu (HFC)$_3$. Under these conditions enantiomeric glycidates form diastereomeric chelat complexes, which give rise
to distinct singulets for C-2 proton and C-3 methyl group of
each of the enantiomers. Essential characteristics of the antipodes are given in table 2. Beside rather low values for optical
rotation especially the difference of odours is noticed, which
should be based on a mechanism of high stereoselectivity.

<u>1Z</u>'   <u>1Z</u>''

20°   20°
[α] = (+) 25,5°   [α] = (-) 26,5°
D   D

c = 15; CCl$_4$ c = 15; CCl$_4$

strawberry-like odour   faint, fruity odour
                                        unspecific

Table 2   Essential characteristics of enantiomers <u>1Z</u>' and <u>1Z</u>''

By reductive cleavage with LiAlH$_4$, 3-phenylbutane-1,3-diol is
received as the only reaction product (10). Dextrorotatory
ethyl glycidate leads to laevorotatory 3-phenylbutane-1,3-diol,
which possesses (S)-configuration as described in the literature (11), and the dextrorotatory reduction product is derived
from laevorotatory ethyl glycidate. Therefore reductive ring
opening reaction is a convenient method to determine absolute
configuration of Z-ethyl glycidate enantiomers (Fig. 5). The
strawberry note is exclusively caused by the 2R,3R-configurated
enantiomer, whereas the odour of the 2S,3S-antipode is faint
and unspecific.

In a variety of fruits, needed for human nutrition, there are
aroma molecules, e.g. olefines, esters, and γ- and δ-lactones as
well as heterocyclic compounds with unknown stereochemistry.
From our point of view research on structure-function relationship of flavour substances will be more and more important.
Revealing these relations is of fundamental scientific interest,

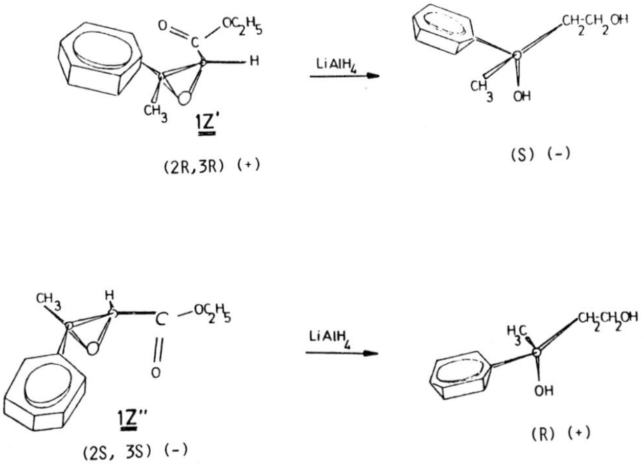

Fig.5   Absolute configuration of 1Z' and 1Z'' enantiomers

and further progress in food analysis should be expected. The German food law differentiates naturally occurring from synthe-sized aromas. The latter ones themselves are subdivided into artificial and so-called "nature identical" flavour substances. In this regard the optical purity of a chiral odorous molecule should be the key to distinguish between "natural occurring" and "nature identical" aromas in food, if optical purity of the natural material is known. In the following, from our current research on naturally occurring aroma compounds the chiral sul-phur-containing flavour substances of passion fruit will be dis-cussed.

Passion Fruit

3-Methylthiohexane-1-ol (3) and 2-methyl-4-propyl-1,3-oxathiane (4) were isolated from the yellow passion fruit and identified by Winter et al. (12). The synthetic route starts from 2-trans-hexenal, and CH$_3$SH, (H$_2$S) is added to yield the saturated C-6-aldehydes, which are reduced by NaBH$_4$ to racemic products. Condensation of 3-thiohexane-1-ol and acetaldehyde leads to cis/trans diastereoisomers of corresponding oxathianes; each of them has to occur as racemic mixture. To get all stereoisomers we made the following modifications:

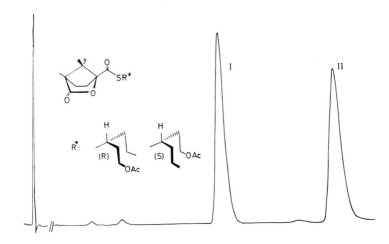

Introducing the acetyl rest as a protective group for primary alcoholic function 91% of monoacetylated product resulted, which was separated in pure form without difficulties. The secondary thiol group of 1b reacted with (1S, 4R) (-) camphanoyl chloride to diastereomeric thiolesters, which were separated by liquid chromatography (petrolether/ethylacetate, 97:3) on silica (Fig. 6).

Fig.6 LC-separation of diastereomeric (1S,4R)(-)camphanoyl thiolesters of (1b)

Reductive cleavage yielded pure enantiomeric thiols of 1a with intensive sulphur notes. The corresponding enantiomeric methyl-thioethers (3',3'') differ in their odour quality: laevorotatory enantiomer shows spice-like, faint odour, whereas the fruity, exotic note is caused by the dextrorotatory enantiomer (Table 3).

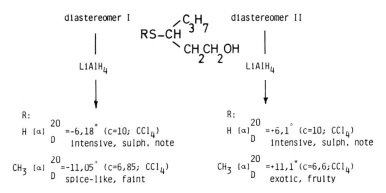

diastereomer I

$$RS-CH \begin{array}{c} C_3H_7 \\ CH_2CH_2OH \end{array}$$

diastereomer II

| LIAlH$_4$ | | LIAlH$_4$ |

R:

H [α]$_D^{20}$ =-6,18° (c=10; CCl$_4$)
intensive, sulph. note

CH$_3$ [α]$_D^{20}$ =-11,05° (c=6,85; CCl$_4$)
spice-like, faint

R:

H [α]$_D^{20}$ =+6,1° (c=10; CCl$_4$)
intensive, sulph. note

CH$_3$ [α]$_D^{20}$ =+11,1° (c=6,6; CCl$_4$)
exotic, fruity

Table 3 Odour qualities of enantiomers of 1a and 3

By cyclisation of optically pure 3-thiohexane-1-ol with acet-aldehyde the expected cis/trans geometric isomers were found. The main product of the 10:1 mixture corresponds to the thermo-dynamically more stable cis-configuration (12).

Each pair of diastereoisomers possessed identical absolute configuration in C-4 position, and we were successful in separating the particular isomers by GC (Fig. 7) and HPLC. Revealing absolute configurations of 3-thiohexane-1-ol enantiomers we referred to the pioneer work of Helmchen et al. (13)on ^{1}H-NMR spectroscopic behaviour of diastereomeric thiolesters derived from 2-phenyl-propionic acid (hydratropic acid HTS). In both diastereomers trans-conformation given by groups -CH-COS-CH is discussed (Fig. 8). This method is based on chemical shift differences comparing protons (of equivalent constitution) of both diastereoisomers. The different shifts for comparable groups result from the up-field shift caused by the

354

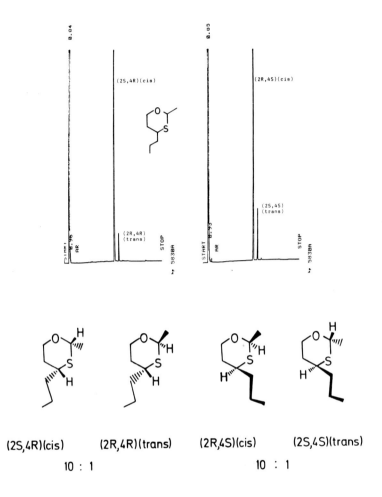

(2S,4R)(cis)   (2R,4R)(trans)   (2R,4S)(cis)   (2S,4S)(trans)

10 : 1   10 : 1

Fig. 7   Capillary GC separation of geometric isomers of 2-methyl-
         4-propyl-1,3-oxathiane (conditions: 25m SE 54; 50°,
         2 min; 5°/min; ret.times: cis-diastereomer, 7.80 min;
         trans-diastereomer, 8.45 min)

phenyl ring. In case of (S)-HTS thiolester diastereoisomers an
up-field shift is expected for the C-4, C-5, C-6 groups in formula A and for C-2, C-1 groups in formula B. In case of (R)-HTS
esters contrary results will be expected. This work is still
under investigation.

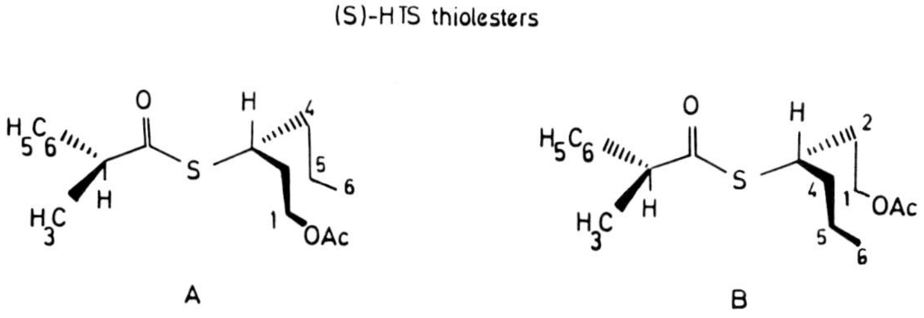

(S)-HTS thiolesters

A   B

Fig. 8 (S)-HTS thiolester diastereoisomers of 1b

If the absolute configuration of 3-thiohexane-1-ol (1a) enantiomers is definitely ascertained, and the preparative separation of oxathiane isomers is finished, full details about
structure-function relationship of these chiral flavour substances of passion fruit will be published (14).

Acknowledgements

This work was supported by the Deutsche Forschungsgemeinschaft.
We are indebted to Prof. Dr. G. Helmchen, Institut für Organi-
sche Chemie, Universität Würzburg, for apparative help and Mrs.
A. Betz and A. Mennig for technical assistance.

References

1. A.J. MacLeod, Symp. Zool. Soc. London 45, 15 (1980).
2. Fenaroli's Handbook of Flavour Ingredients, CRC, The Chemi-
    cal Rubber & Co., Cleveland, Ohio, 1971.
3. M.B. Jacobs, Amer. Perfum. Aromat. 68, 61 and 65 (1956).
4. O. Guillaume, Parfums, Cosmétiques, Arômes 28, 67 (1979).

356

5. A. Mosandl, Habilitationsschrift, Univ. Würzburg 1982.

6. A. Mosandl, Z. Lebensm. Unters.-Forsch. 163, 255 (1977).

7. G. Blaschke, Angew. Chemie 92, 14 (1980).

8. G. Hesse, R. Hagel, Chromatographia 6, 277 (1973).

9. G. Hesse, R. Hagel, Liebig's Ann. Chem. 1976, 996.

10. A. Mosandl, Z. Lebensm. Unters.-Forsch. 177, 129 (1983).

11. S. Mitsui, S. Imaizumi, J. Senda, K. Komo, Chem. Ind. 1964, 233.

12. M. Winter, A. Furrer, B. Willhalm, W. Thommen, Helv. Chim. Acta 59, 1613 (1976).

13. G. Helmchen, R. Schmierer, Angew. Chem. 88, 770 (1976).

14. A. Mosandl, G. Heusinger, in prep.

SIMULTANEOUS DISTILLATION ADSORPTION AND ITS APPLICATION

Hiroshi Sugisawa, Chokou Chen, and Kensuke Nabeta

Department of Food Science, Kagawa University, Mikicho, Kagawa-ken, 761-07, Japan

Introduction

The investigation of quantitative and qualitative differences in volatiles from biological systems, such as callus tissues, insect glands or parts of fruit, requires a convenient, but useful analytical method. In flavour studies, a rapid, but simple and reproducible isolation method of the aroma compounds is essential. The collection of volatiles and the analytical system used to investigate the volatiles should have short sampling times and a minimum of artifacts, respectively. Furthermore, the system should be able to accommodate small sample sizes and be able to isolate and concentrate both low and high boiling compounds.

The problem of isolating the components in complex volatile samples can be achieved by a sequential use of separation techniques in which each step uses different chemical and physical properties to effect separation. Thus, one can use distillation, extraction, adsorption or derivative formation. Adsorption methods have been used for isolation, enrichment and subsequent GC analysis of headspace volatiles in a wide variety of applications (1-4).

In a previous paper (5), we have described a convenient isolation and simple insertion technique for GC, without the interference of solvent. Our simultaneous distillation adsorption method (SDA) was especially developed for small quantiti-

es of biological materials. One disadvantage of this method is
the low recovery of hydrophobic compounds such as terpene hydrocarbons. Since the adsorptive capacity of volatiles by charcoal is much higher with the vapor phase of volatiles than
with volatiles in an aqueous solution, the charcoal method has
been altered to accomplish volatile trapping with the vapor
system.

A new apparatus for the GC analysis of volatiles at the 1 ppm
level, without solvent interference, has been designed and is
described in this paper. The unit combines simultaneous distillation, adsorption and condensation of volatiles. Distilled
volatiles in the vapor phase are trapped in a tube containing
a small quantity of charcoal or porous polymer held between
two stainless steel screens. The condensate returns to the distillation flask for recycling. The method is quantitative
and permits rapid compositional comparisons. It gives good results with samples that are limited in size, because of their
nature or growing stages.

Simultaneous Distillation and Adsorption Apparatus

1) Charcoal trap. Charcoal (GC grade, 30-60 mesh, Wako Chem.
Co.,Japan) was purified with n-hexane and ethyl ether, respectively for 12 hrs each. The purified charcoal was then dried
at 200°C under vacuum for 6 hrs. The drying temperature was
increased gradually to 200°C. The charcoal trap was prepared
by placing ca. 10-50 mg of charcoal in a 4 cm length of 8 mm
OD Pyrex glass tubing with stainless steel screens (120 mesh)
each side of the charcoal layer(5). The trapping tube was connected to the apparatus as shown in Fig. 1.

2) Simultaneous distillation and adsorption. The simultaneous
distillation-adsorption procedure (SDA) for adsorption of distilled volatiles in the vapor phase uses the apparatus shown
in Fig. 1. Before starting a run, the stop-cock is closed, a

To spiral condenser

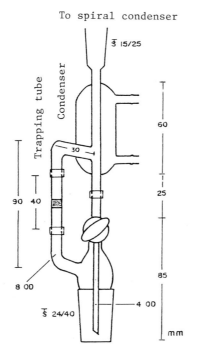

Connected with three-neck flask

Fig. 1   The modified apparatus for simultaneous
distillation adsorption.

three-neck flask (25-100 ml) containing a sample is connected
to the apparatus as well as a steam-inlet and sample is pre-
heated quickly to 60-70°C with a mantle-heater. The run is
then started by introducing steam into the flask.

The SDA was continued for 30 min with the steam. The stop-cock
was used to control flow of the condensate. On completion of
the SDA, the trap was dried for 2 hrs with nitrogen gas, purified by passing it through Molecular sieves 5A and through
Drierite (6) at a flow rate of 40 ml/min.

3)   Desorption of volatiles by solvents. The volatile compounds

trapped onto the charcoal were recovered by solvent extraction. A very small volume (ca. 0.01 ml) of solvent such as
ethyl ether, methylene chloride or Freon 11 (trichlorofluoro-
methane) was added to one end of the trap and a trace of the
eluent was recovered at the opposite end. This was drawn into
a microsyringe and used for the qualitative analysis by GC.
For the quantitative analysis, the charcoal trap was extracted
with 1.0-2.0 ml of solvent. The solution eluted from the charcoal trap was collected into a micro-distillation vessel shown
in Fig. 2, and the solvent removed by hand-warming. The resultant solution (ca. 5-10 μl) was used for GC analysis, with an
internal standard compound.

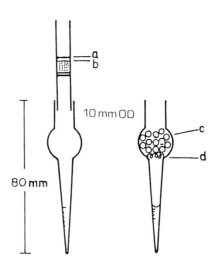

Fig. 2 Micro-distillation vessel.
a: Stainless steel ring, b: stainless
steel screen, c: glass beads, d: stainless steel wire(fine wire).

Table 1   Recovery (%) of model compounds.

| | Total concentration of model solution | | | |
| | 1 ppm | 10 ppm | 1 ppm | 10 ppm |
	SDA-1		SDA-2	
α-Pinene	0 %	25.7 %	90.0 %	99.9 %
Camphene	0	0	96.3	97.7
ß-Pinene	4.8	21.8	96.6	96.6
Carene	0	12.8	93.0	88.1
Ocimene	0	9.9	84.7	92.8
Limonene	0	17.6	94.9	92.7
p-Cymene	0	18.7	95.2	90.0
1-Octene-3-ol	39.3	76.3	98.3	92.4
Camphor	62.7	84.5	95.4	81.5
Linalool	75.6	61.6	93.9	85.8
Linalyl acetate	22.0	40.7	41.1	30.5
ß-Elemene	0	7.4	82.8	60.2
Pulegone	76.9	81.9	37.2	23.9
α-Terpineol	81.2	94.9	74.0	57.3

Recovery (%) relatives to the initial amount of the
model solution.
The values are means from three experiments with SDA
and solvent extraction.

Quantitative GC Analysis

A model solution comprised of 14 monoterpenes with a range of
functional groups and boiling points similar to those commonly
found in plant essential oil was prepared. The components, in
order of elution on the gas chromatogram, are shown in Table
1.   The total concentration of model mixture was kept at 1
ppm or 10 ppm. Thirty ml of the model solution was placed in
a 100 ml flask and the SDA run for exactly 30 min. After the
trapping of volatiles,the charcoal was dried for 2 hrs with
nitrogen gas, as described above. The amount of each component in the mixture adsorbed on the trap was determined on a
Shimadzu GC 6A, with n-tridecane as internal standard. The
fused silica capillary column was 50 m X 0.25 mm and was coated with Carbowax 20M. Column conditions were $70^{\circ}C$ initial followed by $5^{\circ}C$/min to $170^{\circ}C$.

362

Table 1 shows the recovery of the 14 terpenes. Charcoal is not
deactivated by water and adsorbs organic compounds, even from
dilute solutions. However, it is generally known that the adsorptive capacity is decreased in the presence of water or
high humidity (Fig. 3)(7,8). In an aqueous phase, the surface
of charcoal is saturated with water and adsorption of hydrophobic compounds such as hydrocarbons is decreased asymptoti-
cally. Under such conditions, the recoveries of volatile components in an aqueous solution are variable and depend on polarity of the molecule. It is evident in Table 1 that the recoveries of terpene hydrocarbons by the previous SDA apparatus(
SDA-1 in Table 1)(5) are much less than those with the new
SDA method (SDA-2 in Table 1). Most of the compounds were almost quantitatively recovered with a 30 min operation. Recoveries of pulegone and linalyl acetate were somewhat lower, and
about 40% of their initial amounts was still present in the
aqueous solution. Thermal decomposition products were not found in the gas chromatogram.

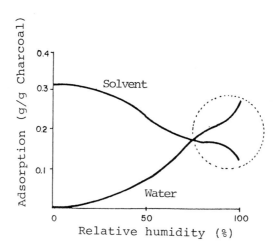

Fig. 3   Effect of relative humidity on adsorption
of organic compounds(7). *Source from Refs. 7 .*

Application of the SDA Method in Volatile Production by Plant
Tissue Cultures

To date, more than twenty papers describe volatile production
by plant tissue cultures. In order to produce commercially
valuable metabolites by this biotechnology, it is necessary to
establish the optimal conditions for both the growth and the
productivity of metabolites by tissues and cells. Selection of
cell variants with higher productivity must be done. To accomplish these needs, it would be desirable to have a facile and
highly sensitive quantitative method for determination of the
metabolites.

In the study of volatiles from tissues or cell suspensions,
sample preparation for GC and GC-MS poses a special problem
because growth of tissues and cells in vitro may be slow and
productivity of volatiles may be low. Separation of volatiles
from callus tissues at 0.01% to 0.001% level of fresh weight
0.5 g to 1.0 g) was adapted to the SDA method. A combination
of simultaneous distillation extraction (SDE)(9), and GC or
GC-MS, resulted in the identification of mono-, sesqui- and
diterpenes, and long chain compounds with a variety of functional groups from three Perilla callus tissues and cell suspensions and from Japanese cedar. Details of the experiment are
described in the following text.

Identification of Volatile Compounds from Callus Tissues and
Cell Suspensions

1) Volatiles from three Perilla strains. Callus tissues and
cell suspensions used for isolation of volatiles were induced
and subcultured as described in previous papers (10-14). The
mono- and sesquiterpenes and long chain compounds which were
produced by three Perilla callus and cell suspensions, Perilla
frutescens Britton var. crispa Decne (Taiwan Aoshiso) and P.

<u>frutescens</u> Britton var. crispa Decne f. purprea Makino (Aka-
chirimen) and <u>P. frutescens</u> Britton var. crispa Decne f. vir-
idae Makino (Aochirimen) are shown in Table 2.

Table 2 Volatile compounds produced by Perilla callus and
cell suspensions.

Compounds	Plants		
	Akachirimen	Aochirimen	Taiwan Aoshiso
Monoterpenes	Limonene α-Pinene ß-Pinene	Terpinen-4-ol	Isoegomaketone Limonene Linalool Perillaketone
Sesquiterpenes	ß-Bisabolene ß-Chamigrene Cuparen Isolongifollene Thujopsene		
Long chain compounds	n-Tridecane n-Pentadecane n-Nonanal n-Decanal 2,4-trans,trans- Decadienal 2,4-trans,cis- Decadienal	Ethylhexanoate Others[a]	
Type of culture	Callus	Callus	Callus and cell suspension
Isolation method	SDA, SDE	SDE	Steam distilla- tion
Identification	GC, GC-MS (EI).[b]	GC, GC-MS (EI,CI).[b]	GC, GC-MS (EI)[b]
References	(11, 13, 14)	(14)	(10, 12)

a) Long chain compounds which possess a common terminal moiety
   gave M^+ ions at 186, 236, 240, 254, 296, 324, 338, 366.
b) EI at 70 eV , CI at 200 eV with isobutane.

In the volatiles from Akachirimen, the main constituents were hydrocarbons, n-alkanes, mono- and sesquiterpenes. Most of sesquiterpenes in the callus tissues were not detectable in the essential oil obtained from intact leaves. Long chain aldehydes were also observed in the callus. On the contrary, Aochirimen callus did not give a variety of mono- and sesquiterpenes. Terpinene-4-ol was found with other unique long chain compounds in the volatiles. The unique long chain compounds had MS spectra with fairly intense molecular ion peaks and characteristic base ions at m/z 59. The base peak indicated the existence of $Me_2C(OH)-$, EtCH(OH)- or MeCH(OMe) moiety in the molecules, but the former two moieties were eliminated by the fact that 1,1-dimethyl-9-octadecen-1-ol and 1-ethyl-9-octadecen-1-ol were not strong molecular ion peaks but more like dehydrated ions. Methyl-1-methyl-9-octadecenyl ether, which was chemically synthesized from oleic acid in several steps, gave observable molecular ion and base ion peaks at m/z 59. The latter strongly suggested the existence of MeCH(OMe) moiety in the homologous compounds. Final confirmation of the structures of these compounds are in progress. Isolation procedures, means of identification and references are also shown in Table 2.

2) <u>Volatiles from the cell suspensions of cedar</u>. The characteristic volatiles from these suspensions were diterpenes which are also present in the intact plant (15). Fatty acids, esters, and aldehydes listed in Table 3 were also identified. Mono-and sesquiterpenes were not detected in the volatiles. Diterpenes identified were ferruginol and abietatriene (Fig. 4).

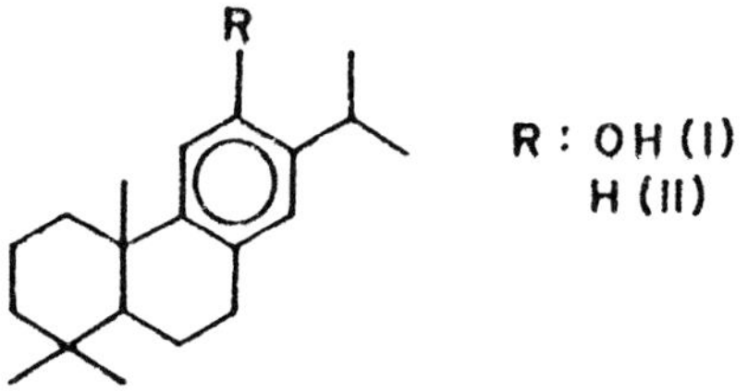

R: OH (I)
H (II)

Fig. 4. Ferruginol (I) and abietatriene (II).

366

Table 3 Long chain compounds in the volatiles from
cell suspensions of Japanese cedar
(Cryptomeria japonica (15)).

Compounds	
Aldehydes	Two 2,4-heptadienals
	n-Nonanal
	2,4-Nonadienal
	Two 2,4-decadienals
Esters	Methyl palmitate
	Ethyl palmitate
	Methyl linoleate
	Methyl linolenate
	Ethyl linolenate
Acids	Myristic acid
	Palmitic acid

3) <u>Limitations and prospects for the SDA method</u>. When the SDA
method is used for the isolation of volatiles from callus tissues, the main problem will be the adsorption characteristics
of adsorbents for the volatile compounds. In a previous paper
(5), we mentioned that the recoveries of mono- and sesquiterpenes by SDA (SDA-1) with charcoal varied from 8% to 100%. Differences between the SDA-1 method and the SDE method are shown
in Fig. 5. The example shows the volatiles isolated from the
cell suspensions of Japanese cedar (Cryptomeria japonica) by
SDA (SDA-1, Fig. 5a) and SDE (Fig. 5b). In the chromatogram of
SDA, peaks 15 and 23 were absent, or present in very small amount. Thus, in order to use the SDA method for quantitative
analysis, one should identify the volatiles by means of GC-MS,
NMR or IR to ascertain the volatiles.

In our laboratory, SDA was routinely used in the study of volatile production by plant tissue cultures, particularly in the
relationship of the growth of callus tissues as influenced by
growth regulators and precursors to production of volatiles.
In each trial, samples with volatiles at 10 µg to 100 µg level

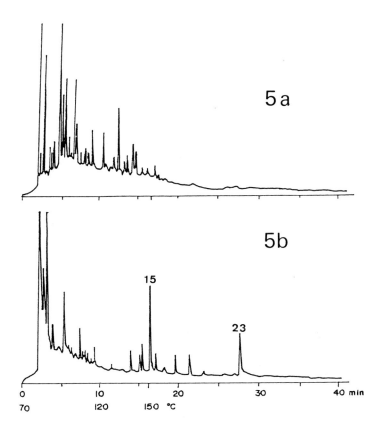

Fig. 5   Gas chromatograms of volatiles from cell
          suspensions of Japanese cedar.
          5a: SDA-1, 5b: SDE. Column: 30 m X 0.3 mm
          WCOT glass capillary coated with Carbowax
          20 M. Column temp. 70-150°C at 5°C/min.
          Flow rate N_2 0.78 ml/min.

were analyzed daily. By the use of SDA method, we ascertained
that the volatile production was synchronous with the growth
of callus tissues from Akachirimen. Furthermore, the addition
of precursors such as acetate or mevalonic acid markedly enhanced volatile production.

Summary

For the study of volatiles from small quantities of biological material, a rapid, facile and reproducible isolation method is required. A new apparatus for GC of volatiles in quantities of 0.5 g to 1.0 g, free from solvent interference, has been developed.The unit combines simultaneous distillation adsorption and condensation. The adsorption of distilled volatiles is done in the vapor phase by charcoal trapping, and the condensate returns to a flask for recycling. The method is quantitative and permits rapid compositional comparisons. Reliable and reproducible results from callus tissues and cell suspensions were obtained with the new method.

Acknowledgement

The authors wish to thank Dr. R. Teranishi, Western Regional Research Center, United States Department of Agriculture, Berkeley, California, for a gift of stainless steel screen.

References

1. Clark, R.G.,Cronin, D.A.: J. Sci. Fd. Agric. 26, 1615 (1975).

2. Jennings, W.G., Filsoof, M.: J. Agric. Food Chem. 25,440-445 (1977).

3. Murray, K.E.: J. Chromatogr. 135, 49-60 (1977).

4. Klimes, I., Stunzi, W., Lamparsky, D.: J. Chromatogr. 136, 13-21 (1977).

5. Sugisawa, H., Hirose, T.: Flavour'81; Schreier, P. ed.; p. 287-299, Walter de Gruyter, Berlin 1981.

6. Sugisawa, H.: Flavor Research, Recent Advances; Teranishi, R., Flath, R.A., Sugisawa, H. ed.; p.11-51, Marcel Dekker, New York 1981.

7. Mizutani, M., Nagano, S.: Active Carbon, p.205-218, Kodansha, Tokyo 1975.

8. Kalab, P.: Collect Czech. Chem. Commun. (CSK) 47, 2491-

2500 (1982).

9. Schultz, T.H., Flath, R.A., Mon, T.R., Eggling, S.B., Te-
    ranishi, R.: J. Agric. Food Chem. 25, 446-449 (1977).

10. Sugisawa, H., Ohnishi, Y.: Agric. Biol. Chem. 40, 231-232
    (1976).

11. Nabeta, K., Sugisawa, H.: Plant Tissue Culture '82; Fuji-
    wara, A. ed.; p.289, Japan Assoc. Plant Tissue Culture,
    Tokyo 1982.

12. Nabeta, K., Ohnishi, Y., Hirose, T., Sugisawa, H.: Phyto-
    chem. 22, 423-425 (1983).

13. Nabeta, K., Kataoka, M., Hirose, T., Sugisawa, H.: Agric.
    Biol. Chem., submitted for publication.

14. Nabeta, K., Sugisawa, H.: Insrumental Analysis of Foods,
    Recent Progress, vol.1; Charalambous, G., Inglett, G.ed.;
    p.65-84, Academic Press, New York 1983.

15. Ishikura, N., Nabeta, K., Sugisawa, H.: 8th Plant Tissue
    Culture Symp. Abstr. p 65, Toyama, Japan, July 1983.

A "CLOSED-LOOP-STRIPPING" TECHNIQUE AS A VERSATILE TOOL FOR METABOLIC STUDIES OF VOLATILES

Wilhelm Boland, Peter Ney and Lothar Jaenicke
Institut für Biochemie, An der Bottmühle 2
D-5000 Köln 1, F.R.G.

Günter Gassmann
Biologische Anstalt Helgoland, Meeresstation
D-2192 Helgoland, F.R.G.

Introduction

The metabolism of compounds in living organisms is commonly followed by monitoring the isotope flux of specifically labelled precursors. This procedure is well established and useful for gaseous products. In the case of lipophilic metabolites, however, the damage or even destruction of intact cells by organic solvents sets limits and prevents continuous long-term tracing in a single specimen in vivo. Also the pre-chromatographic purification of crude extracts results in non-predictable effects which might confuse the final answer.

Therefore, a mild and non-destructive method of removal of volatiles from living organisms combined with a suitable sample treatment and enrichment of compounds is highly desirable.

In recent years, considerable efforts have been devoted to analyse for volatiles by immediate sampling techniques such as CO_2 extraction (1), cold trap condensation (2) and - particularly - head-space procedures (3,4,5). Of the latter the "Closed-Loop-Stripping" technique of Grob and Zürcher (6) has proven the most efficient. Originally developed for air-water stripping, its extension to air-solid

extraction promises to be excellently suited for
investigating metabolizing living systems over longer
periods of time under natural conditions.

Since only air circulating in a closed system is used to
blow off the volatile products and to transport them to an
adsorbent charcoal pad, no permanent disturbance (beside
shifts in equilibrium) of metabolic turnover in the
organism(s) has to be expected.

As an application of this methodology a simple device is
presented by which volatile secondary metabolites from
labelled fatty acids fed to intact plants are collected and
analyzed.

Materials and methods

The basic concept of our stripping device corresponds to the
original description of Grob and Zürcher (6). The air volume
is circulated continuously through a closed system by means
of a membrane pump (a, Metal Bellows Corp., USA). It passes a
small charcoal filter (b, Brechbühler, Switzerland) which
retains the the volatile constituents. The all glass filter
housing (c, Otto Fritz GmbH, Normag; Gassmann et al.,
unpublished) consists of a heater (low voltage power supply)
and a high precision glass tube which holds the removable
charcoal filter tightly in place. The filter holder and pump
may be attached to various vessels depending on the sample
size. In our case, 100 to 500 mL round bottom flasks for
leaves and seedlings or small animals and 2.5 L flasks for
whole plants were used.

Constituents of the lower boiling point range were
stripped out during the first 4 hrs while those of the higher
boiling point range required up to 24 hrs to be extracted. If
only a screening of volatiles is the aim the (plant) material
may be chopped to increase surface and yield. After stripping

the compounds were desorbed from the filter with 30 μL dichloromethane, and aliquots were subjected to GC and GC-MS analysis.

A Finnigan (quadrupole) mass spectrometer, model 4510 GC/MS, equipped with a DB 5 fused silica capillary (30m x 0.31 i.D.; J & W) was used for isotope analysis. The ionization potential was 70 eV, and the scan range was from 35 to 400 amu/sec. Scanning and data processing were accomplished with an INCOS data system. Selective ion monitoring was applied for the detection of labelled metabolites.

Figure 1. Schematic drawing of the stripping apparatus

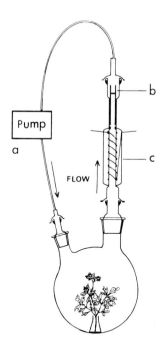

a) Stainless steel membrane pump, model MB-41-E, Metal Bellows Corp., Sharon, Mass., U.S.A.

b) Precision glass liner containing 1.5 mg of activated carbon embedded between gold plated steel screens, Brechbühler AG, CH-8952 Schlieren/ZH, Switzerland.

c) All glass filter housing with an integrated heating coil, Otto Fritz GmbH, Normag, D-6238 Hofheim/Ts., F.R.G.

Rotulex bowl joints (Sovirel) with fused glas-metal transitions served as gas-tight, flexible connections between the pump and the stripping apparatus.

Results and discussion

The mild and non-destructive characteristics of this sampling technique have been shown by the distinct stripping of volatile constituents from different types of living organisms: germinating seedlings of <u>Carthamus</u> <u>tinctorius</u>, stems of <u>Senecio</u> <u>isatideus</u> and of live bumble bees.

Figures 2a,b and 5 show the GC profiles of volatiles from all these organisms.

The chromatogramm of the germinating seedlings (Fig. 2a) shows more than 100 different substances ranging from the boiling point of n-octane up to eicosane. Heptadeca-1,8,11,14-tetraene 1, heptadeca-1,8,11-triene 2 and 1-penta-decene 3 are the most prominent compounds. During the first stripping intervals (less than 8 hrs) preferentially the low boiling components are trapped while subsequent longer stripping enriches the less volatile C_{17}-polyenes. Time course and extracted amounts are compiled in Figure 3.

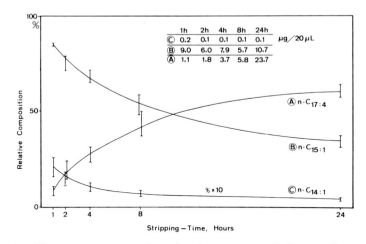

Figure 3. Time course and relative composition of extracts. Absolute amounts were determined with external standards and refer to the total extracts (20 µL).

When <u>Carthamus</u> seedlings are grown in water containing double-bond deuterium labelled pentadeca-6,9,12-trienoic

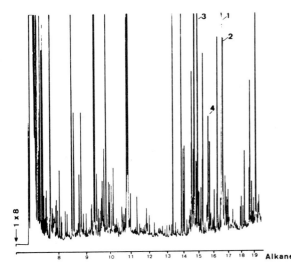

Figure 2a

Volatile constituents from germinating seeds of Carthamus tinctorius.

Conditions:
30 seedlings;
72 hrs stripping

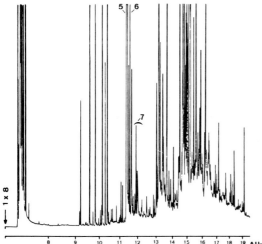

Figure 2b

Flavour components from a cutting of Senecio isatideus (Compositae) in water.

Conditions:
4.5 g plant material;
12 hrs stripping

Chromatograms of plant strippings by an on-column injection technique on CP-SIL-5, 50m x 0.32mm (Chrompack, Netherlands), under programmed conditions from 25° (5 min) to 220° at 5°/min; H_2 = 0.6 bar; FID 1 x 8, 220°; sample size 1 μL
The elution order of n-alkanes is given instead of retention times to allow classification of compounds.

acid, tetradeca-1,5,8,11-tetraene and hexadeca-1,7,10,13-tetraene 4 can be stripped off and prove this particular fatty acid as the immediate precursor (equ. 1).

$$\text{n-}C_{15:3} \longrightarrow \text{n-}C_{14:4} + \text{n-}C_{16:4} \qquad \text{equ. 1}$$

$$3\text{-OH-}C_{17:3} \longrightarrow \text{n-}C_{16:4} + \text{n-}C_{18:4} \qquad \text{equ. 2}$$

Feeding of 3-hydroxy-heptadeca-8,11,14-trienoic acid gives a similar result. Again two new olefines bearing a deuterium label are formed. Mass spectroscopy and GC-comparison with authentic references indicate hexadecatetraene 4 and octadeca-1,9,12,15-tetraene to be present (equ. 2).

We have to conclude from these results that vinyl-group formation from fatty acids proceeds via 3-hydroxy intermediates which are fragmented heterolytically by enzyme catalyzed loss of carbon dioxide and water (7). If the administered fatty acid is elongated by a two-carbon unit prior to the final fragmentation step, the concomittant formation of higher molecular weight olefines is easily explained in the same line.

The Senecio chromatogram (Fig. 2b) reveals a number of C_{11}-hydrocarbons, among which ectocarpene 5, S-(+)-6-(1Z-butenyl)-1,4-cycloheptadiene, is the most abundant. Others are 6-butyl-1,4-cycloheptadiene 6 and several undecatetraenes 7 (cf. Fig. 2b). Ectocarpene 5 is also found as a gamete attractant of marine brown algae (8). Although no experimental data on its biosynthesis are accessible to date, the structural elements of this and other algal pheromones suggest fatty acids as the possible precursors (8).

To test for this hypothesis, dissolved tritium labelled

oleic-, linoleic- and linolenic acids were fed to cuttings of the plant in water, and the volatiles analyzed by our method and by conventional extraction techniques. In both cases no incorporation into ectocarpene could be observed. However, 30 hrs feeding of dodeca-3,6,9-trienoic acid (tritium labelled) resulted in significant radioactivity in the extracts. Their gaschromatographic and radioactivity profile, as shown in Fig. 4, proves emergence of ectocarpene 5 and radioactivity in the same fraction.

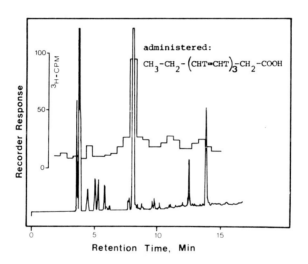

Figure 4. Gaschromatographic and radioactivity profile of a *Senecio* stripping after incubation with ^{3}H-dodecatrienoic acid. Compounds were separated on a CB-WAX 57 CB fused silica column (50m x 0.32mm, Chrompack, Netherlands). The column effluent was analyzed with a photo ionization detector and collected in 30 sec intervals. Liquid scintillation counting was used to monitor radioacticity.

This indicates that the pool of multiply unsaturated medium chain fatty acids contains the precursor for ectocarpene (equ. 3). The unnatural undeca-3,6,9-trienoic acid (3,4-6,7,9,10-deuterium labelled) was also taken up and a 6-(1Z-propenyl)-1,4-cycloheptadiene, still containing all deuterium markers, was formed (molecular ion: m/z=140). Also deuterium in the 5-position of this acid remained unaffected,

378

thus excluding this particular methylene group as being directly involved in the activation step (equ. 4).

COOH $\longrightarrow$ equ. 3

S-(+)-Ectocarpene

COOH $\longrightarrow$ equ. 4

equ. 5

COOH $\longrightarrow$

Other experiments, using deca- or trideca-3,6,9-trienoic acid (R=H, C_3H_7) as possible precursors resulted in very low conversions (equ. 4). Only minute amounts of vinyl- or pentenyl-1,4-cycloheptadiene were detectable by GC-MS. This may point to a high specificity of the active site of the proposed converting enzyme(s).

When the plant is grown in the presence of dodeca-3,6-dienoic acid, butyl-1,4-cycloheptadiene 6 (found in another algal species as a pheromone (8)) is the only reaction product (equ. 5). Though the precursor is obviously synthesized by the plant itself (cf. Fig. 2b), isotope analysis revealed more than 50 % de novo synthesis from the externally offered, deuterated fatty acid. These results suggest that indeed the same basic enzymatic mechanisms are responsible for the production of various pheromone-homologues.

The experiments presented demonstrate how fast and reliable metabolic questions can be answered by this simple and direct gas-solid stripping method.

As a final demonstration, living bumble bees were caught from the institute garden and seven individuals were enclosed for 4 hrs into a 2.5 L stripping flask. The result is given in Fig. 5. More than 100 substances could be separated. Two major products were identified as dodecanoic acid ethyl ester 8 and farnesol 9 while the third component 10 proved to be a terpene-ketone of unknown structure.

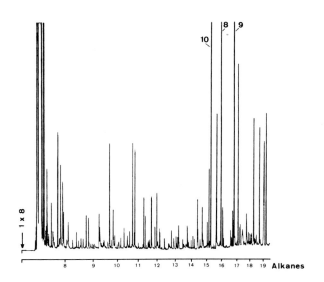

Figure 5. Volatile constituents from bumble bees.
Seven individuals were enclosed in a 2.5 L circulation flask and stripped at ambient temperature (25° C). All insects kept alive and could be released after four hours. Instrumentation and analytical conditions were as described in Figure 2.

The last stripping experiment extends the range of organisms to living animals and secures that there is no principal limitation if adequate environmental conditions can be maintained.

References

1. Jennings, W.G., in: Recent Advances in Capillary Gas Chromatography, Volume 2, (Bertsch, W., Jennings, W.G., Kaiser, R.E., EDS.) pp. 195-205, Dr. Alfred Hüthig Verlag, Heidelberg · Basel · New York 1982

2. Redshaw, E.S., Hougen, F.W., Baker, R.J.: J. Agric. Food Chem. 19 (6), 1264-1271 (1971)

3. Michael, L.C., Erickson, M.D., Parks, S.P., Pellizari, E.D. Anal. Chem. 52, 1836-1841 (1980)

4. Withycombe, D.A., Mookherjee, B.D., Hrouza, A., in: Analysis of Food and Beverages: Headspace Techniques, (Charalambous, G., Ed.), Academic Press, New York 1978

5. Dupuy, H.P., Brown, M.L., Legendre, M.G., Wadsworth, J.I., Rayner, E.T., in: Lipids as a Source of Flavour, (Supran, K.M., Ed.), pp. 60-67, ACS Symposium Series, Washington, D.C. 1978

6. Grob, K., Zürcher, F.: J. Chromatogr. 117, 285-294 (1976)

7. Bohlmann, F.: Fortschr. Chem. Org. Naturst. 25, 1-50 (1968)

8. Jaenicke, L., Boland, W.: Angew. Chem. 94, 659-669 (1982); Angew. Chem. Int. Ed. 21, 643-653 (1982)

25°C and equilibration time of 30 minutes were kept constant. After elution of adsorbed volatiles with ether and addition of internal standard substances, the eluate was concentrated and separated in an unpolar and a polar fraction by column chromatography on silica gel. The compounds were identified by mass spectral analysis and headspace concentrations were determined by analyzing concentrates on a 25 m CW 20 M glass capillary column (8).

Thus with a detection sensitivity of about 100 ng/l a headspace concentration range, respectively a mean value for each component in headspace could be ascertained, representing the approximate quantitative proportions as perceived by the nose. According to the morphological differentiation of lemon and other citrus fruits, peel and fruit, as well as juice were separated prior to the headspace analysis. Determining influence of interactions of volatiles with different lemon tissues on headspace concentrations, a cold pressed lemon peel oil (Aromachemie, Aufsess) and a water-peel oil emulsion of the peels, separated by pressing, were additionally analyzed. Table 4 shows headspace concentration ranges and mean values of some selected aroma compounds in different lemon samples.

The terpene hydrocarbons, especially limonene are the main constituents in the headspace of all samples. The main polar compounds in cold pressed lemon peel oil and the water-peel oil emulsion, separated from the peel are citral, 1- octanal and 1-nonanal, while nerol, geraniol and n-alcanols are dominating in the headspace of lemon peel pieces. Concentrations of peel oil aroma compounds in the headspace of peel-oil-free lemon juice are considerably minor, while compounds of high volatility, not yet identified, are dominating (8). These results illustrate the well-known influence of the sample matrix on the headspace concentrations of the volatiles (9,10).

Table 3 Recognition threshold ranges and odour qualities of stimuli of some lemon aroma compounds, eluted from the sniffing port stimulus elution time corresponding to the peak width : 5 sec.
threshold ranges : nanogram / stimulus
       N   : number of single threshold determinations

aroma compound	threshold range	N	odour quality
1-octanal	0.5 – 0.9	50	lemon–fruity,aldehyde
ethylhexanoate	0.6 – 1.0	20	fruity, rum-like
1-undecanal	0.9 – 1.2	26	aldehyde
1-dodecanal	1.4 – 2.0	27	aldehyde
1-hexanal	2 – 4	50	green
1-decanal	3 – 4	23	orange–fruity,aldehyde
(+)-linalool	3.5 – 4.0	15	flowery, with fruity note
neral	3 – 5	20	lemon
1,8 cineole	3 – 5	20	medicinal,eucalyptus
1-nonanal	4.5 – 6.0	26	lemon–fruity,aldehyde
geranial	5.5 – 7.0	20	lemon
citronellal	5.0 – 6.5	15	aldehyde, soapy
myrcene	13 – 15	15	like hops
geraniol	15 – 20	32	flowery
citronellol	80 – 100	25	flowery
limonene	220 – 230	41	terpeny, fruity–sweet
nerol	320 – 370	37	flowery,with fruity note
geranylacetate	1200 – 1800	16	fruity
nerylacetate	2000 – 2500	15	citrus-fruity
α-pinene	2500 – 2900	22	terpeny,turpentine-like
ß-pinene	3500 – 3800	45	coniferous
α-terpineol	10000 – 11000	15	flowery

3.   The study of flavour formation in apples by treating in-
     tact apples with volatile precursors.

For trapping headspace volatiles cryogenic or adsorptive methods can be used. Nowadays the use of porous polymers or
small amounts of activated carbon is widespread. Although in
the past the emphasis in aroma research was on the identification of odorous compounds, for practical applications (e.g.
objective-subjective correlations) attention should be given
to quantitative aspects. An important question is whether the
adsorption, the desorption and the manipulations required to
transfer the sample to the chromatograph are quantitative. In
this study we use Tenax GC for trapping volatiles, liberated
from intact fruits or during disintegration of fruits. The
adsorbed components are thermally desorbed and concentrated in
a cold trap, allowing injection on high resolution capillary
columns. The described thermo desorption and cold trap injection system is self-constructed. High resolution gas chromatography is required for separation of the complex aroma patterns (3). Component identification is performed by GC-MS
analyses.

Materials and Methods

Dynamic headspace sampling during fruit disintegration. The
sampling apparatus for representative isolation of volatiles,
released during fruit disintegration, is presented in Figure
1. (see also reference 4). A commercial blender A (Braun MX
32, Germany), in which all plastic parts had been changed by
Teflon, was adapted with a glass flange B and fitted with a
three-necked flange cover. The cover was fitted with a helium
supply through C and a splash head D with dump valve E, which
was connected to the adsorber F.

The adsorbent trap F consisted of a relatively large glass tube (i.d. 1.6 cm; length 10 cm) packed with 5 g of Tenax GC 60/ 80 mesh. The helium gas flow, regulated by a fine metering valve, was measured by the flow meter G and the total volume of the gas sample by the wet-testmeter H.

For dynamic headspace sampling about 250 g of fruits were used. Before starting fruit disintegration, the inside space of the blender was washed with helium through dump valve E during 10 minutes. For trapping headspace volatiles the blender was flushed with helium at a flow rate of 500 ml/min, while disintegrating the fruits slowly (the rotation speed of the propeller was regulated by a Vareac). Adsorption was continued for 10 min and the total gas volume over the adsorber was 5 L. Under the conditions employed adsorption was quantitative and no breakthrough occured for products with a retention index (on SE-30) higher then 700. This was checked by sampling on two connected adsorption tubes, which were separately analyzed. An attractive feature of this sampling procedure is that it is clearly related to the sensory perception. Fresh fruits are macerated during consumption, which certainly has an influence on the release of volatiles and may lead to instant formation of certain volatiles as a result of enzymatic reactions. However for study of the evolution of the aroma composition (e.g. as a function of the degree of ripening) and for the study of flavour formation headspace sampling of intact fruits has important advantages (5).

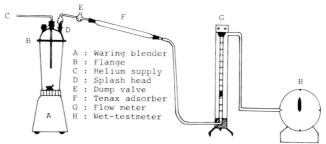

A : Waring blender
B : Flange
C : Helium supply
D : Splash head
E : Dump valve
F : Tenax adsorber
G : Flow meter
H : Wet-testmeter

Figure 1 Apparatus for isolation of volatiles, released during fruit disintegration.

Dynamic Headspace sampling of intact fruits. For dynamic headspace sampling without fruit disintegration, apples (1.5-2 kg)
were placed in 8 L desiccators (thermostated at 18°C), which
were continuously flushed with air at a flow rate of 150 ml/min.
Before sampling, the air flow was increased to 400 ml/min, and
after 30 min a Tenax GC 60/80 mesh tube (length 11 cm; i.d.
12 mm; 1.8 g adsorbent) was attached to the outlet of the desiccator and sampling was continued for 15 min. after re-adjusting
the air flow to 400 ml/min. Loaded adsorption tubes, when
tightly closed and kept in the refrigerator, could be stored
for a long period without loss of sample or changes in the composition.

For the flavour formation studies volatile precursors (acetic,
propanoic and butanoic acids, propanal, butanal) were added by
installing a 2 necked flask containing 100 µl of the product
between the air pressure tank and the desiccator as presented
in Figure 2. The compounds evaporated at ambient temperature
and were absorbed by the fruits. Less volatile compounds
(pentanoic and hexanoic acids; pentanal and hexanal) were directly injected into the desiccators.

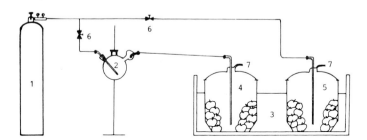

Figure 2   Apparatus for headspace collection from intact apples
            and for treatment of apples with volatile precursors.
            1 : high purity purge gas, 2 : evaporation flask for
            additives, 3 : thermostated waterbath (18°C), 4 : des-
            iccator with treated fruits, 5 : desiccator with un-
            treated fruits (blank), 6 : fine metering valves,
            7 : outlet for connection of adsorption tubes.

Thermo desorption and cold trap injection. Sample recovery
was performed by heat desorption and collection in a cold trap,
which allowed a sharp injection on high resolution capillary
columns. For this purpose the gas chromatograph was modified
with a self-constructed injection system, consisting of a
desorption oven and a two-position six-port, high temperature
injection valve (Valco Instruments Co., Houston, Tx.). For
sample injection the glass adsorption tube was placed in a
separate oven (gradually warmed up to 220°C) and connected
to the six-port, high temperature injection valve by a glass
adaptor, a reducing union, 1/16 in. tubing and thermally isolated zero-volume fittings. The injection valve, which was
also connected to the capillary column, was constantly kept
at 250°C. In transfer position the flavour components were
desorbed in a helium stream (transfer gas) and collected in
a trap cooled with liquid nitrogen. When desorption was complete, the 6-port injection valve was switched from collection position (trap flushing with transfer gas) to injection
position (trap flushing with carrier gas) and the cold trap
was heated with a flood lamp of 1250 W, performing an instan-
taneous injection into the capillary column.

Gas chromatography and mass spectrometry. The volatiles from
the different strawberry samples were separated on a Varian
3700 gas chromatograph, equipped with a 200 m x 0.6 mm i.d.
glass column coated with SE-30 and an effluent splitter for
simultanuous flame ionisation and flame photometric detection. Further operating conditions were as follows : linear
temperature programming from 20 to 220°C at 2°C/min.; carrier gas He, 3 mL/min.; injection valve temperature 250°C;
detector temperature 250°C. The apple volatiles were separated on a 150 m x 0.5 mm i.d. glass column, coated with
SE-52. Linear temperature programming was from 20°C to 200°C
at 1°C/min.

This example evidently shows the influence of headspace concentrations on the aroma significance of single components.
Factors influencing volatility like a complex matrix or
different phase states (9,10) can only be included into the
calculations of aroma values based on threshold values in
air and headspace concentrations (11).

Aroma values of aldehydes in the headspace of peel-oil-free
lemon juice are in the range of about 5 to 20 (fig.7).
Evaluating juice headspace concentrate with the gas chromatographic sniffing test, some compounds of high volatility,
i.e. showing short retention times on the FFAP-column (fig.4),
developed fruity odours of relative strong intensity; they
are not yet identified.
For a complete characterization of lemon aroma, these compounds have to be analyzed quantitatively too. However, reproducible analysis of compounds of high volatility by means
of solvent extraction of adsorbing tubes is very troublesome.
Disadvantages of this method are e.g. high adsorption volumes
required, losses during concentrating or the solvent peak.
Thus thermal desorption in combination with column switching
techniques seems to be the best method for quantitative headspace analysis of compounds with high volatility.

Table 6 shows some aroma compounds with very low recognition
threshold values of their pulse-type stimuli. Especially
those of some ethylesters are very low, so for example ethyl
2-methylbutanoate and ethylisopentanoate, with recognition
thresholds of about 10 picogram, eluting from the sniffing
port within 5 seconds.
The ethylesters, also characterized by typical odour qualities
play an important role in many fruit flavours. By calculations
of odour units in water, Guadagni and coworkers (12) identified ethyl 2-methyl-butanoate as an "character impact compound"
of "golden delicious" apple essence. Von Sydow and coworkers
(18) evaluated the same compound to be important for bilberry

aroma. Finally Schultz and coworkers (19)identified this ester
as well as ethylbutanoate and ethylisobutanoate in the headspace of peel-oil-free orange juice. Assuming headspace concentrations of 100 ng / l, aroma values based on these sniffing threshold values would be in the order of 1000 for ethyl
2-methylbutanoate, 200 for ethylisobutanoate,and 50 for ethyl
butanoate. These examples illustrate that trace compounds of
high volatility can have a high aroma effectivity in fruits.

Table 6 Volatiles with low odour recognition threshold ranges
of stimuli eluted from the sniffing port; stimulus elution
time corresponding to the peak width : 5 sec.; threshold ranges : nanogram / stimulus

compound	threshold range	odour quality
ethyl 2-methyl-butanoate	0.006 - 0.012	"golden delicious" apple
ethylisopentanoate	0.007 - 0.01	medicinal
ethylisobutanoate	0.03 - 0.05	fruity,orange-like
ethylpentanoate	0.03 - 0.04	fruity,pineapple-like
ethylbutanoate	0.2 - 0.3	fruity,rancid note
ethanethiol	0.08 - 0.09	cabbage-like
1-octanol	0.5 - 0.9	lemon-fruity, aldehyde

Structure-activity relationships

Finally, odour recognition threshold values determined with
the gas chromatographic sniffing technique can be a basis for
a discussion of structure-activity relationships in human
chemoreception.
In table 7 some examples of different threshold values of
stereo and geometrical isomers are listed like (+)-linalool

and n.hexanal), which showed important fluctuations dependent on the degree of maturity and were formed enzymatically during disintegration, should contribute to the strawberry flavour. Presentation of the complete quantitative data is not possible due to extensiveness. As an illustration in Figure 3 the chromatogram of the variety Sivetta together with the identified compounds, numbered according their programmed temperature retention indices, are presented. It can be observed that the esters are the qualitatively and quantitatively most important volatiles in strawberry flavour. The sensory relevant

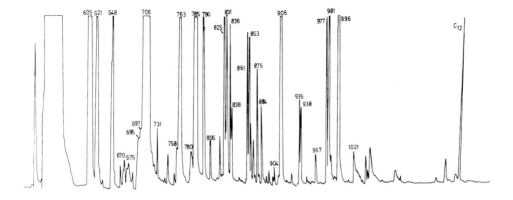

Figure 3   Chromatogram of strawberry variety Sivetta : 200 m x 0.6 mm i.d. SE-30 capillary, linear temperature programming 20°C → 220°C, FID-detection, peak numbering by means of retention indices.
605 = ethyl acetate, 621 = methyl propionate, 648 = isopropyl acetate, 670 = methyl 2-methylpropionate, 670 = 2-pentanon, 675 = methylthiol acetate, 695 = ethyl propionate, 697 = propyl acetate, 706 = methyl butyrate, 731 = dimethyl disulfide, 758 = 2-methylpropyl acetate, 763 = methyl 2-methylbutyrate, 780 = n-hexanal, 785 = ethyl butyrate, 796 = butyl acetate, 806 = methyl pentanoate, 825 = isopropyl butyrate, 831 = trans-2-hexenal, 836 = ethyl 2-methylbutyrate, 838 = ethyl 3-methylbutyrate, 861 = 3-methylbutyl acetate, 863 = 2-methylbutyl acetate, 875 = methyl 4-methylpentanoate, 884 = propyl butyrate, 904 = isopent-2-enyl acetate, 906 = methyl hexanoate, 938 = 2-methylpropyl 2-methylpropionate, 967 = 4-methylpentyl acetate, 977 = butyl butyrate, 981 = ethyl hexanoate, 996 = hexyl acetate, 1021 = isopropyl hexanoate.

sulfur compounds could only be distinguished in some varieties and are frequently hidden by the major constituents or by compounds with analoguous chromatographic properties.

At a latter stage of this research project (from the seasons 1980 up to now) when the importance of the minor sulfur constituents had been recognized, strawberry varieties were evaluated for flavour quality using multiple detection gas chromatography. As sulfur compounds can be detected specifically by flame photometric detection, an effluent splitter was installed for simultanuous FID- and FPD-detection. For comparison of varieties the complete method consisted of headspace enrichment on Tenax (1 L headspace) during fruit disintegration, thermo desorption-cold trap injection, high resolution gas chromatography and quantitation by normalisation of the values from electronic integration against an external standard solution, containing 1 µg tridecane and 200 ng dipropyl disulfide. As an illustration the chromatogram with FID- and FPD-detection of the variety Gorella and Souvenir de Charles Machiroux are presented in Figure 4. Table 1 presents quantitative data for sulfur compounds expressed as ng S/1 L of headspace). It was clear that varieties (e.g. Souvenir) which were indicated by the taste panel as significantly more strawberry flavour intensive, had a more complete and intensive pattern of sulfur containing compounds.

As a conclusion we may state that correlating GC data with taste panel results is a convenient way for determination of relevant flavour compounds. Herewith use can be made of wide flavour quality differences, which can be demonstrated by sensory assessment of e.g. different varieties. Furthermore the importance of a representative isolation of volatiles should be stressed.

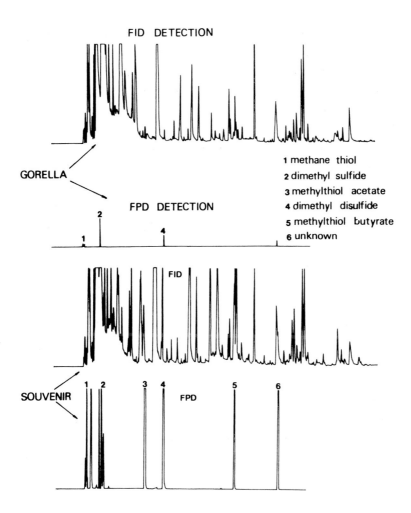

FID DETECTION

GORELLA

FPD DETECTION

1 methane thiol
2 dimethyl sulfide
3 methylthiol acetate
4 dimethyl disulfide
5 methylthiol butyrate
6 unknown

SOUVENIR

FID

FPD

Figure 4   GC-analyses with simultaneous flame ionisation- and
            flame photometric detection of the headspace vola-
            tiles of the strawberry varieties Gorella and Sou-
            venir.

Objective measurement of aroma quality in Golden Delicious
apples. As a result of consumer complaints concerning a decreasing flavour quality of fresh fruits a growing interest
is observed for objective measurement techniques. Indeed,
sensory analysis is not suitable for evaluation of flavour

392

Table 1   Quantitative data for sulfur compounds, expressed as
ng S/1 L of headspace, for different strawberry varieties.

Varieties	MeSH	EtSH	MeSMe	MeCOSMe	MeSSMe	BuCOSMe	Un-known	Sum
Primella	19	26	19	–	21	–	–	85
Tioga	25	42	72	31	180	30	–	380
Early Bommel	45	61	49	31	99	18	–	303
Gorella	22	–	76	–	6	–	–	104
Hapil	29	–	161	–	12	–	–	202
Korona	12	13	112	–	13	–	–	150
Red Gauntlet	26	–	68	–	79	–	9	182
Sivetta	33	6	102	17	177	8	19	362
Souvenir	56	54	158	116	529	88	122	1123

influencing parameters, as it is extremely difficult to compare sensory quality of products, which are not available
at the same moment (e.g. evolution of flavour quality as a
function of storage time in controlled atmosphere). Furthermore sensory analysis is always a momentary evaluation,
while flavour quality in fruits is a dynamic process, which
is importantly influenced by the degree of ripening.

Flavour in fruits is the combined effect of their constituents on the taste (non-volatiles : sugars and acids) and
olfactory organs (volatiles). As aroma is one of the key
factors in fruit flavour quality, it can be used as a criterion for evaluation of flavour influencing parameters.
Recently we studied the influence of controlled atmosphere
storage on apple flavour, using headspace enrichment of volatiles released during fruit disintegration (9). In the
study discussed here, emphasis was on the influence of
picking date and storage time in controlled atmosphere on
the flavour quality of the economically important Golden
Delicious apples (70% of the Belgian production). In order
to follow the evolution of the volatile composition as a

function of ripening a non-destructive headspace sampling
of intact apples represented several advantages, such as
convenience and increased reproducibility, because the same fruits were sampled throughout each ripening experiment.
GC patterns of apple volatiles by headspace sampling with
or without disintegration were rather similar, but some
differences were observed mainly as a result of enzymatic
formation (hexanal and trans-2-hexenal) or degradation (lower concentration of high boiling esters due to esterase
activity) of compounds during fruit disintegration. The importance of enzymatic reactions during fruit processing is also
discussed by Schreier (10).

The flavour character of various apple cultivars can differ
markedly according to their volatile composition (11). However, flavour differences between apples of the same cultivar should be mainly due to quantitative differences in their
composition of aroma and taste constituents. In table 2 the
evolution of Golden Delicious volatiles (isolated by headspace sampling of the intact fruits) as a function of ripening in standard conditions immediately after harvesting
(picking date : October 8, 1980) is presented. The concentrations in Table 2 are expressed as micrograms per 6 L of
headspace per Kg of apple and were obtained by normalisation
of the peak intensities from an electronic integrator against
n-undecane and n-tridecane as internal standards. It is
shown that the Golden Delicious aroma is composed of an important amount of volatile esters, some alcohols, estragole
(4-methoxyallylbenzene) and farnesene. These compounds are
only produced in higher amounts during ripening, related to
the climacteric rise in respiration and are only present in
small concentrations during growth or at the time of harvest.
From Table 2 it is also clear that a complex patterns of
carboxylic esters is responsible for the fruity character
of apple flavour and as a criterion for objective flavour
quality measurement the total sum of esters could be used.

Table 2   Relative amounts (expressed as micrograms per 6 L of headspace per Kg of apple) of Golden Delicious volatiles as a function of ripening in standard condition immediately after harvesting (picking date : October 8, 1980).

Compound	Days of ripening							
	1	5	8	12	14	16	19	22
Ethyl acetate	-	-	-	0,08	0,09	0,10	0,12	0,15
Butanol	0,03	0,11	1,22	4,17	5,14	4,80	6,38	7,25
Propyl acetate	-	0,01	0,10	1,31	2,45	4,10	5,23	6,67
2- of 3-methylbutanol	-	-	-	0,20	0,43	0,34	-	0,62
2-methylpropyl acetate	0,01	0,04	0,16	0,35	0,34	0,30	0,32	0,23
Propyl propionate	-	-	-	-	-	-	-	0,31
Butyl acetate	0,53	1,73	15,35	64,97	81,45	78,25	79,41	71,12
3-methylbutyl acetate	0,22	0,30	1,52	9,83	13,38	12,78	12,59	11,53
Hexanol	-	-	0,77	1,74	1,70	1,13	3,09	3,00
Propyl butyrate	-	-	-	0,09	0,30	0,22	0,25	0,33
Butyl propionate	0,02	0,05	0,69	3,58	5,01	5,21	4,74	5,50
Pentyl acetate	0,05	0,16	0,78	2,10	3,26	2,25	2,15	1,92
Propyl 2-methylbutyrate	-	-	0,03	0,03	0,09	0,11	0,07	0,10
Butyl butyrate	0,09	0,42	2,51	5,45	5,70	4,39	3,42	3,12
Hexyl acetate	0,45	1,95	17,68	60,65	78,11	68,20	68,87	62,89
Butyl 2-methylbutyrate	-	0,08	0,78	2,58	2,76	2,06	1,81	1,36
Pentyl butyrate Butyl pentanoate	-	0,21	0,27	1,13	0,87	0,82	1,07	0,90
Hexyl propionate	0,03	0,14	1,18	4,51	8,27	7,21	8,10	8,75
Heptyl acetate	-	0,02	0,02	0,05	0,10	0,10	0,18	0,09
Pentyl 2-methylbutyrate	-	-	0,04	0,10	0,09	0,08	0,06	0,06
Hexyl 2-methylpropionate	-	0,04	0,09	0,11	0,13	0,09	0,13	0,05
Butyl hexanoate Hexyl butyrate	0,30	1,57	9,28	15,13	20,91	15,58	15,49	14,87
Estragole	-	-	0,19	0,64	1,00	0,63	0,77	0,72
Octyl acetate	-	-	0,03	0,04	0,04	0,05	0,05	0,05
Hexyl 2-methylbutyrate	0,06	0,15	2,36	8,65	13,27	9,65	10,89	9,81
3-methylbutyl hexanoate	-	-	-	0,01	0,03	0,03	0,01	0,03
Pentyl hexanoate Hexyl pentanoate	-	0,01	0,16	0,33	0,56	0,42	0,33	0,51
Heptyl 2-methylbutyrate	-	-	-	0,04	0,03	0,09	0,05	0,06
Hexyl hexanoate	0,18	0,23	2,28	4,31	6,45	5,25	6,31	6,59
Farnesene	0,23	0,88	12,32	27,75	42,59	30,38	33,82	39,97
Sum esters	1,94	7,11	57,39	185,43	243,69	217,34	221,75	207,00

Figure 5 shows the important influence of picking date on apple aroma. The evolution of the sum of esters is presented as a function of days of ripening (in standard conditions) for different picking dates : premature picking : September 3, 1980; early picking : September 23, 1980; medium picking date : October 10, 1980 and late picking : October 27, 1980. By following the evolution of volatiles as a function of ripening a complete picture of the dynamic flavour quality process is given. Late picked apples show an immediate increase of volatiles and reach a considerably higher maximum compared to early picked apples.

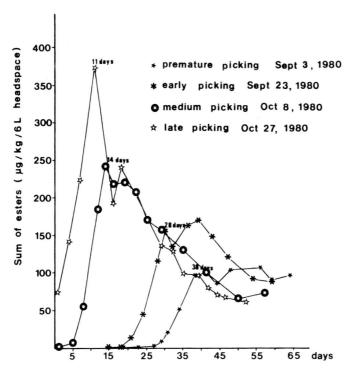

Figure 5   Influence of picking date on the evolution of the sum of Golden Delicious esters (expressed as µg per Kg per 6 L of headspace) as a function of days of ripening in standard conditions immediately after harvesting.

As nowadays a considerable part of the apple production is
stored in controlled atmosphere (CA) it was of major importance to evaluate the influence of CA storage on apple flavour. In Figure 6 the evolution of the total sum of esters
as a function of ripening in standard conditionsafter removal from controlled atmosphere is given for early, medium
and late picked apples. Ripening experiments were performed
after short (until January 12, 1981), medium (until March 10,
1981) and long (until May 25, 1981) CA storage. In relation
to eating quality it is accepted that consumption is about
10 days after removal from controlled atmosphere (distribution,
shop and home storage). From examination of Figure 6 two important considerations can be concluded : a) there is a marked
decrease of aroma after long CA storage for all picking dates;
b) only the medium picking date shows a maximum after long CA
storage. From the considerations above it may be concluded
that in order to offer better quality regulations for shorter
storage should be established. Otherwise criteria should be
worked out for determination and prediction of the optimum
picking date for apples intended for CA storage.

Flavour formation in apples by treating intact apples with
volatile precursors. Several reviews have gathered information on the biogenesis of aroma constituents in fruits and
it was stated that more research should be devoted to the elucidation of the process of aroma formation and degradation in
fruits (12,13,14). In our laboratory up to now attention was
given to the biogenesis of carboxylic esters, as chief constituents in apples. Usually incorporation experiments in
apples were performed by treating simplified systems such
as apple discs (15) or cell suspension (16) with carboxylic
acids, alcohols and aldehydes. However flavour formation
studies can be performed on intact apples, as they absorb
readily volatile substances when kept in an atmosphere containing these additives. Furthermore dynamic headspace
sampling allows following the evolution in function of time

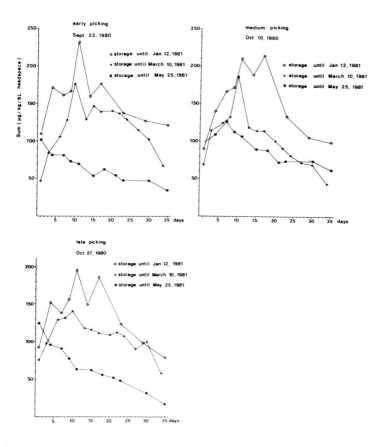

Figure 6   Evolution of the sum of Golden Delicious esters as a function of ripening after removal from controlled atmosphere (CA storage). Influence of picking date and storage time.

of the volatiles, produced by intact apples after treatment with volatile precursors.

Up to now Golden Delicious apples were treated with C_3- to C_6-aldehydes and C_2- to C_6-carboxylic acids. In all cases these additions had a marked influence on the volatile pattern and on the total volatile production. A full discussion of the effect on the headspace composition (as a function of ripening

398

in standard condition) of the treatment of Golden Delicious
apples with aldehydes and carboxylic acids is given by De
Pooter et al. (17).

As an illustration of the effect of adding carboxylic acids,
Figure 7 presents a graphical picture of the volatiles, isolated by dynamic headspace sampling 1 day after treatment
with acetic, propanoic and butanoic acids in comparison with
the composition of untreated apples. It is shown that added
carboxylic acids were transformed into esters or alcohols
(or smaller carboxylic acid by β-oxidation where possible),
which are then used by substrate volatiles for ester formation. Important qualitative changes in the volatile pattern

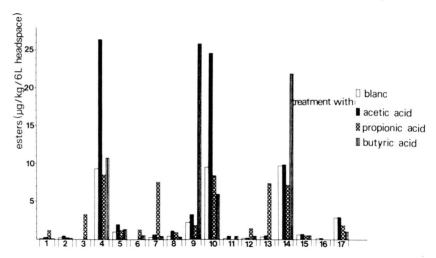

Figure 7   Composition of Golden Delicious headspace 1 day af-
            ter treatment with acetic, propionic and butyric acid.
            Comparison with the composition of untreated apples.
            1=propyl acetate, 2=2-methylpropyl acetate, 3=pro-
            pyl propionate, 4=butyl acetate, 5=3-methylbutyl
            acetate, 6=propyl butyrate, 7=butyl propionate, 8=
            pentyl acetate, 9=butyl butyrate, 10=hexyl acetate,
            11=butyl 2-methylbutyrate, 12=sum of propyl hexa-
            noate, butyl pentanoate and pentyl butyrate, 13=
            hexyl propionate, 14=sum of butyl hexanoate and
            hexyl butyrate, 15=hexyl 2-methyl butyrate, 16=
            hexyl pentanoate, 17=hexyl hexanoate.

are observed after adding propanoic acid, as propanoates and
propylesters are absent or present in only small concentration
in the blank.

Summary

This contribution wants to demonstrate the possibilities of
dynamic headspace sampling for different applications in
fruit flavour analysis. It is shown that this technique
has important advantages over classical sample preparation
techniques for determination of relevant flavour compounds,
for objective measurement of flavour quality and for flavour formation studies. If may be stated that the described applications would not have been possible with conventional isolation procedures.

References

1) Charalambous, G. : Analysis of Foods and Beverages. Headspace techniques. Academic Press, New York - San Francisco
   - London 1978.

2) Kolb, B. : Applied Headspace Gas Chromatography. Heyden,
   London - Philadelphia - Rheime 1980.

3) Jennings, W. : Gas Chromatography with Glass Capillary
   Columns. Second edition, Academic Press, New York -
   London - Toronto - Sydney - San Francisco 1980.

4) Dirinck, P., Schreyen, L., Schamp, N. : J. Agric. Food
   Chem. 25, 759 (1977).

5) De Pooter, H., Dirinck, P., Willaert, G., Schamp, N. :
   Phytochemistry 20, 2135 (1981).

6) Nursten, H. in : The Biochemistry of Fruits and their
   Products (Hulme, A.C., Ed.). Vol I, p. 242, Academic
   Press, London 1970.

7) Van Straten, S., de Vrijer, F.L., de Beauveser, J.C.
   (Eds.) in : Volatile Components in Food Nutrition and
   Food Research. TNO, Zeist, The Netherlands.

8) Dirinck, P.J., De Pooter, H.L., Willaert, G.A., Schamp,
   N.M. : J. Agric. Food Chem. 29, 316 (1981).

9) Willaert, G.A., Dirinck, P.J., De Pooter, H.L., Schamp, N.M. : J. Agric. Food Chem. 31, 809 (1983).

10) Schreier, P. in : Quality in Stored and Processed Vegetables and Fruit (Goodenough, P.W., and Atkin, R.K., Eds), pp. 355-371, Academic Press, London - New York - Toronto - Sydney - San Francisco 1981.

11) Williams, A.A. in : Progress in Flavour Research (Land, D.G. and Nursten, H.E., Eds.), pp. 287-305, Applied Science Publishers Ltd., London 1978.

12) Tressl, R. and Drawert, F. : J. Agric. Food Chem. 21, 560 (1973).

13) Salunkhe, D.K. and Do, J.G. : Crit. Revs. Food Sci. Nutrit. 8, 161 (1977).

14) Eriksson, C.E. in : Progress in Flavour Research (Land, D.G. and Nursten, H.E., Eds.), pp.159-174, Applied Science Publishers Ltd., London 1978.

15) Paillard, N. : Phytochemistry 18, 1165 (1979).

16) Ambid, C. and Fallot, J. : Bull. Soc. Chim. Fr. II, 104.

17) De Pooter, H.L., Montens, J.P., Willaert, G.A., Dirinck, P.J., Schamp, N.M. : J. Agric. Food Chem. 31, 813 (1983).

HEADSPACE ANALYSIS FOR THE STUDY OF AROMA COMPOUNDS IN MILK AND DAIRY
PRODUCTS

H.T. Badings and C. de Jong, NIZO, Ede, the Netherlands

1. *Introduction*

The use of instrumental analytical techniques for the rapid evaluation
of the flavour quality of raw milk can be of great help to determine
whether this milk is suitable for further processing and preparation of
certain products. Changes in flavour quality of processed milk and milk
products as a result of microbial, enzymic or chemical deterioration may
also be monitored by such techniques.

For routine use these methods of analysis must be simple,
straightforward and rapid to perform. Headspace analysis by gas
chromatography (GC) meets these demands. However, sampling a few
millilitres of equilibrated headspace vapours from a closed vial
containing the product, by means of a gas-tight syringe, will give
satisfactory results only if the aroma compounds have a sufficiently
high vapour pressure and a flavour threshold value which is not too
low. The method can be greatly improved as regards sensitivity and
repeatability, by stripping the volatiles from the product with a stream
of inert gas and collecting the entrained compounds on a pre-column or
in a cold trap (purge-and-trap method). By heating the trap, the
volatiles are subsequently transferred to a GC column for further
analysis. An example of a simple on-column trapping procedure for GC
analysis of flavour volatiles, is the technique described by Morgan &
Day (1). Headspace or purge-and-trap procedures have been used for the
analysis of aroma compounds from milk and milk products by Palo et al.
(2, 3, 4), by Dumont & Adda (5) for UHT milk, by Marsili (6) for
cultured butter milk and by Wellnitz et al. (7) for detection of fruity
flavours in milk.

In the present work a purge-and-trap system is described for

Analysis of Volatiles
© 1984 Walter de Gruyter & Co., Berlin · New York – Printed in Germany

GC/headspace analysis of volatiles from aqueous and powdered samples. The entrained compounds are collected in a capillary cold trap. This trap is subsequently heated in order to transfer the volatiles to a capillary GC column where they are separated directly. The whole configuration can be operated semi-automatically.

The first series of experiments with the designed system has been focussed on off-flavours in (raw) milk as a result of microbial spoiling, heating, or exposure to light.

2. Experimental

2.1. Headspace gas chromatography

The system designed for purge-and-trap-headspace GC (PTHGC) is given schematically in Figure 1. The purge-and-trap unit consists of a 50 ml sample flask ① provided with a piece of glass tubing ending at the bottom of the flask in a glass frit which is so designed that a flow of finely dispersed purge gas (helium) is led through the sample. The gas stream then passes a cold trap ②, held at -10 °C by means of alcohol circulated by a cryostat (Landa TK 300). In this trap the surplus of water vapour is collected, which otherwise would block the second cold trap ④. In this second trap, made of glass-lined tubing (length 10 cm, 1,0 mm inner diameter) and cooled at approx. -130 °C (cooling with liquid N_2 using a Cryoson cryogenic unit), the volatile components from the sample are collected. During the purge-and-trap operation, gas flow ⑨a (9 ml He/min) is used and valve ③ is held in position I. After having passed trap ④, the purge gas flows through the injector port ⑩ of the GC into the capillary column.

Further details of the purge-and-trap operation are: sample of milk 40 ml, temperature 30 °C, purge flow 9.0 ml/min through the sample flask (due to back pressure, the flow through the capillary column starts at 2 ml/min and increases to 9 ml/min in 13 min); purge time 20 min.

For GC analysis of the collected volatiles, valve ③ is turned to position II. This reduces the flow through the capillary column to 5.0 ml/min within 2 min. Then the cold trap is heated by applying a direct current to the GLT tubing (4 V) as a result of which its temperature is raised to approximately 250 °C within 2 min.

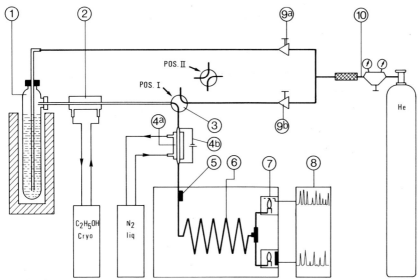

Figure 1. System for purge-and-trap headspace gas chromatography.

1. Sample flask with constant-temperature mantle.

2. Cold trap, cooled by ethanol from a cryostat.

3. Two-way valve: position I = purge-and-trap operation; position II = GC analysis.

4. Glass-lined tubing, either cooled with liquid N_2 (4a = trapping), or heated by direct current (4b = evaporation).

5. On-column injector.

6. Capillary column ending in a 1:1 T-splitter.

7. Detectors (FID and FPD).

8. Recording and data-processing.

9. Flow-adjustment: 9a = purge-and-trap flow, 9b = GC analysis flow.

10. Helium gas supply with filters for dust, water and oxygen.

For further details see text.

The heat-released volatiles are transferred through the injection port (220 °C) onto the capillary column, 75 m x 0.5 mm inner diameter, which is coated with OV-1/Carbowax 20 M (1:1), d_f = 1.2 µm, and held at 40 °C for isothermal analysis. The end of the column is connected to a T-splitter which transfers the carrier gas - in a 1:1 ratio - to a flame ionisation detector (FID) and a flame photometric detector (FPD)(7),

both held at 160 °C. The attenuation range for the FID is x 10^{-12}, and
that for the FPD is x 10^{-7}.

The whole set of steps in the purge-and-trap procedure and the GC
analysis is controlled by a Perkin Elmer Sigma 10B data system⑧which
also calculates the results of the GC-FID/FPD analyses. Only the setting
of valve③is not automated as yet.

An extra on-column injector⑤(8) is installed in the oven at the
beginning of the capillary column. This injector is used to compare the
performance of the system in direct injection with that of the system in
purge-and-trap mode.

2.2. Calibration of the PTHGC system and analysis of samples
The following components were added to milk (fat content 3 %) and to
water in concentrations of 0-100 µg/kg: dimethyl sulphide, dimethyl
disulphide and a mixture of 2- and 3-methyl butanal. The samples for
calibration were prepared just before PTHGC analysis. They were poured
into a stoppered sample flask, after which the purge-and-trap operation
was started. Foaming of the milk, if any, was controlled by adding a few
mg of 1-tetradecanol. Quantitation was done by integration of the peak
area.

2.3. Sensory evaluations
A panel of six milk graders experienced in the evaluation of milk
flavours participated in the experiments. On the one hand samples were
evaluated to determine the nature and the intensity of the (off)flavours
(scale: 0 = absent, 1 = slight, 2 = moderate, 3 = strong, 4 = very
strong). On the other hand the panel assisted in determining flavour
threshold values for certain off-flavour compounds in milk.

3. Results and discussion

3.1. Performance of the PTHGC system
In Figure 2 two examples are given of the performance of the analytical
system. Figure 2A shows the results of a headspace analysis of an
aqueous sample containing a number of aroma compounds. The difference in
resolution between direct injection of the compounds and PTHGC analysis

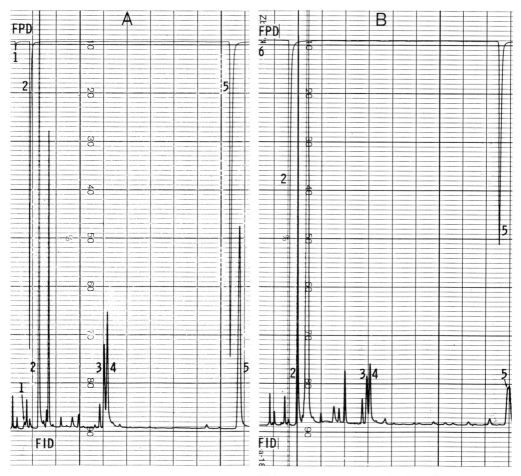

Figure 2. Results obtained with the PTHGC system. 2A: analysis of an
aqueous sample containing a number of aroma compounds in µg/kg
quantities; 2B: sample of milk with the same aroma compounds added.
1 = methyl mercaptan; 2 = dimethyl sulphide; 3 = 3-methyl butanal;
4 = 2-methyl butanal; 5 = dimethyl disulphide.
Response is given for FID and FPD detector.

was negligible. In Figure 2B results are given of a sample of milk with
added aroma compounds. Volatile compounds from the milk itself causes
additional peaks. These peaks do not interfere with the peaks of the
added aroma compounds, with one exception: dimethyl disulphide.

Fortunately the additional compound is not detected in the FPD recording, because it does not contain sulphur.

Collection of water in trap ② and of volatiles in trap ④ was found to be satisfactory. During the purge-and-trap operation no disturbance of the detector base line due to break-through of components was observed. A blank run between two analyses also gave a clean base line, which demonstrates that there is no hold-up of aroma compounds in the cooled trap ② or in trap ④ after heating. Only in the case of high concentrations of volatile compounds (> 50 mg/kg) was some memory effect observed.

3.2. Calibration curves

The calibration curves of a number of aroma compounds are given in Figures 3, 4 and 5. Because of the non-linear response characteristic of the flame photometric detector (9), the calibration curves made with this detector (Figs. 3B and 4B) are given on a semi-logarithmic scale. The response appears to be lower in milk than in water, because the presence of milk fat (3 to 4 % in whole milk) and/or other constituents may reduce the vapour pressure of the volatile compounds.

From Figure 2 it can be seen that the two isomers of isopentanal which are present (2-methyl butanal and 3-methyl butanal) are well separated by the present method of analysis. However, since the ratio of the one isomer to the other in the reference material used for calibration is different from that of these compounds in deteriorated milk, it was necessary to give the results of quantitative analysis as the sum of the two isomers.

It should be noted that for the calibrations in milk a blank run was made in order to make correction for the volatiles that were originally present.

3.3. Repeatability

The repeatability of the PTHGC system was tested in experiments with a number of aroma compounds at a level of approx. 20 µg/kg in water. The results, together with the minimum detection levels, are given in Table 1. At this low concentration the percentage of variation was between 8 and 20, which was acceptable in view of the object pursued:

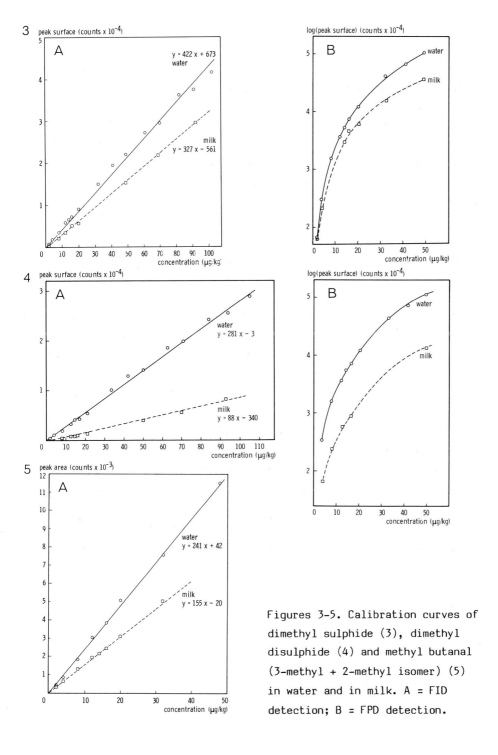

Figures 3-5. Calibration curves of dimethyl sulphide (3), dimethyl disulphide (4) and methyl butanal (3-methyl + 2-methyl isomer) (5) in water and in milk. A = FID detection; B = FPD detection.

Table 1. Standard deviation of the PTHGC determination of a number of
         volatile compounds and minimum quantity which can be determined

compound	method[1]	number of expt. (n)	results (µg/kg) average (x)	results (µg/kg) s.d. (s)	variation (%)	minimum quantity which can be determined
dimethyl sulphide	FID	9	20.0	1.6	8	1
dimethyl sulphide	FPD	7	20.5	3.8	19	2
dimethyl disulphide	FID	9	21.0	2.5	12	5
dimethyl disulphide	FPD	7	22.0	4.4	20	4
methyl butanal	FID	9	22.0	2.9	13	1

[1] Method of detection: FID = flame ionisation detector, FPD = flame
photometric detector in sulphur mode.

the estimation of levels of aroma compounds in relation to the
occurrence of certain off-flavours.

It should be pointed out here that the highest percentages of
variation given in Table 1 (19 and 20) are not caused by errors in the
PTHGC system, but are due to instability of the FPD in hand.

3.4. Recovery of added compounds

Table 2 gives the results of the analysis of a number of volatile
compounds added to milk (3 % fat) in concentrations between 10 and
80 µg/kg. Correction was made for volatiles originally present in the
milk.

3.5. Analysis of milk samples with flavour defects

In order to test the usefulness of the PTHGC system, a number of milk
samples with normal flavours or with off-flavours were studied by
instrumental analysis and by sensory evaluations. It was of particular
interest to compare the results of these two methods. Results of a study

Table 2. Recovery of volatile compounds added to milk.

compound added		results of PTHGC (quantity in µg/kg)	
name	quantity (µg/kg)	FID	FPD
dimethyl sulphide	20	21.5	22
dimethyl sulphide	40	43	37
dimethyl sulphide	80	89	76
dimethyl disulphide	20	17	16
dimethyl disulphide	40	37	37
dimethyl disulphide	60	54	50
isopentanal	10	11	-
isopentanal	20	22	-
isopentanal	40	39	-

Table 3. Results of the PTHGC analysis of raw milk samples with a normal
flavour and of samples with malty or unclean flavours resulting
from microbial spoilage.

milk samples			compounds (µg/kg milk)[3]			
No.	defect	intensity[1]	methane thiol	dimethyl sulphide	methyl butanal	methyl disulphide
1	malty	2	-	5 (0.3)	73 (5)	-
2	malty	3	-	7 (0.4)	187 (12.5)	-
3	unclean	1-2	-	20 (1.5)	-	-
4	unclean	3	-	90 (50.6)	-	-
5	normal[2]	-	-	5-10 (0.3-0.6)	1 (0.07)	-

[1] Scale for intensity: see text on sensory evaluations.

[2] Result of analyses of more than 20 samples of raw and pasteurized
milk without flavour defects.

[3] First figure: quantity of compound in µg/kg milk; second figure:
between brackets, quantity of compound divided by flavour threshold
concentration.

on microbially induced off-flavours in milk are given in Table 3.

It is known that growth of Streptococcus lactis, var. maltigenes may cause malty off-flavours as a result of the production of aroma compounds (10, 11). In particular 3-methyl and 2-methyl butanal contribute to this flavour defect. We estimated the flavour threshold for the sum of these compounds in milk at approximately 15 µg/kg.

Sample 1 (Table 3) had a malty off-flavour of moderate intensity. Indeed the concentration of methyl butanal in this sample was found to be approximately five times its flavour threshold concentration. Sample 2 had a strong malty flavour. The concentration of methyl butanal in this sample was 12.5 times its flavour threshold value.

Unclean flavours in milk are the result of increased concentrations of dimethyl sulphide (11). Although this compound may contribute to a desirable raw-milk flavour when present in low concentrations (5-10 µg/kg), it will cause unclean flavours if a level of 16 µg per kg milk is exceeded. In accordance with this observation, sample 3 (Table 3), which contained 20 µg dimethyl sulphide/kg of milk, had a slight unclean flavour. Sample 4, with a strong unclean flavour, contained 90 µg dimethyl sulphide/kg of milk.

Table 4. Results of PTHGC analyses of raw milk which deteriorated during storage at 9 °C (quantities in µg/kg milk).

	storage time (days)		
	0	1	4
flavour	normal	slightly unclean	unclean, malty, spoilt
compounds			
methane thiol	–	–	15
dimethyl sulphide	5	16	40
ethane thiol[1)	–	–	15
methyl butanal	–	–	57
dimethyl disulphide	–	–	–

[1) tentative

As a reference, Table 3 also gives the levels of dimethyl sulphide and methyl butanal found in samples of milk with a normal flavour.

Table 4 summarizes the results of an experiment in which raw milk was subjected to microbial spoilage during storage at 9 °C for some days. It can be seen that the initial milk had a normal flavour and a sub-threshold concentration of dimethyl sulphide. After one day the milk

Table 5. Results of PTHGC analyses of samples of pasteurized, UHT treatedor sterilized milk (quantities in µg/kg of milk).

compounds	low pasteurized milk (12 s, 73 °C)	UHT milk (32 s, 142 °C)	sterilized milk A[1]	B[1]
hydrogen sulphide	0.5	50	–	–
methane thiol	–	tr	5	5
dimethyl sulphide	7	8	3	18
methyl butanal	1	3	1	3
dimethyl disulphide	–	4	4	4

[1] In-bottle sterilized milk. Further process conditions not available

Table 6. Results of the PTHGC analyses of milk exposed to sunlight (quantities in µg/kg of milk).

compounds	time of exposure to sunlight (h) 0	1	2
acetaldehyde[1]	15	150	250
methane thiol	–	trace	2
dimethyl sulphide	2	2.5	2
acetone[1]	600	550	550
methyl butanal	1	1	4
dimethyl disulphide	–	trace	4

[1] Estimated on the basis of calibration curves for water. The acetaldehyde peak may have been affected by minor quantities of contaminants.

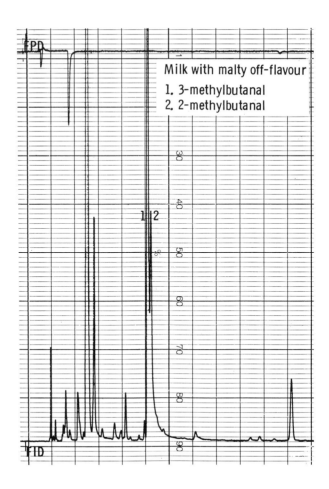

Figure 6. Milk with malty off-flavour.

became slightly unclean, which agrees with the detection of an increased
concentration of dimethyl sulphide having reached the flavour threshold
value. After four days of storage, the milk had a distinct unclean,
malty and spoilt off-flavour. In line with this, unacceptable
concentrations of methane thiol, dimethyl sulphide, ethane thiol and
methyl butanal were found. Dimethyl disulphide was not detected.

In another experiment attention was paid to aroma compounds
contributing to heat-induced flavours. Table 5 gives some results

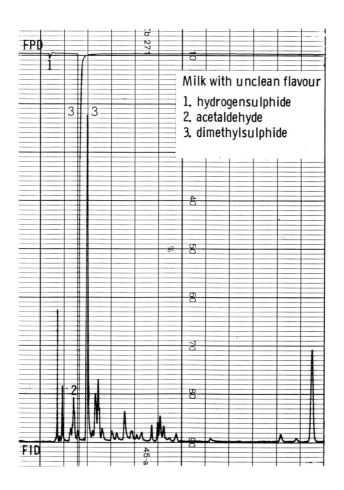

Figure 7. Milk with unclean flavour.

concerned with pasteurized, UHT and sterilized milk. The flavour of
low-pasteurized milk comes close to that of raw milk. Heat-induced
flavour compounds were indeed not detected. In UHT milk, an increase in
the quantity of hydrogen sulphide, methane thiol and methyl butanal,
was observed. A similar observation was made in sterilized milk. The
concentration of dimethyl sulphide showed a wider variation. The results
obtained are - generally speaking - similar to those found by Dumont &
Adda (5). However, in contrast to the latter results, the present work

414

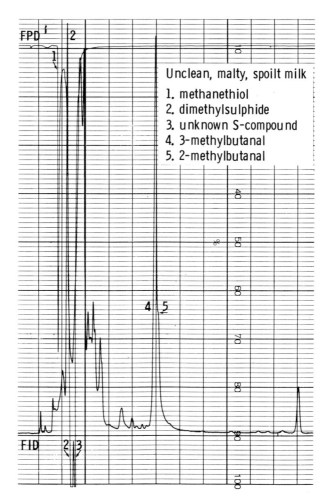

Figure 8. Spoilt milk with unclean, malty flavour.

indicates absence of hydrogen sulphide from conventionally sterilized milk.

Results of an experiment on light-induced flavours in milk are given in Table 6. Some of the compounds listed were also determined by Bassette (12). The results of these investigations and the present work are in fair agreement with each other, although there is some difference in concentration. The concentration of acetaldehyde rises sharply when milk is exposed to light. However, the concentrations of dimethyl

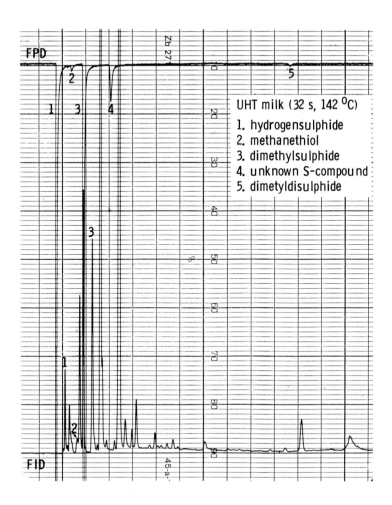

Figure 9. UHT milk (32 s, 142 °C).

sulphide and acetone did not change very much. In addition, the
concentrations of methane thiol, methyl butanal and dimethyl disulphide
were found to have increased significantly after milk was exposed to
light. A number of these compounds may contribute to light-induced
flavours in milk (13). Yet their concentration is very low and comes
close to the detection limit of the present method of analysis.

In conclusion it can be said that the present study has demonstrated
the usefulness of a PTHGC system for the analysis of those volatile

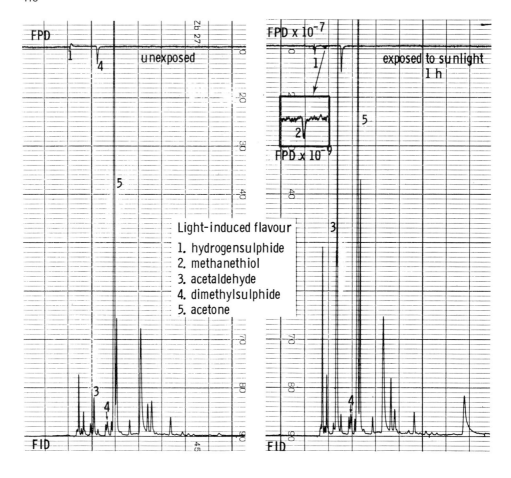

Figure 10. Milk with light-induced flavour.

compounds which contribute to desirable or undesirable flavours in milk. The examples given show that the results of instrumental analytical determinations of volatile aroma compounds can be used either to monitor the flavour quality of milk or to confirm the results of sensory evaluations. As an illustration, Figure 6 shows the results of PTHGC analysis of a sample of milk with a malty off-flavour. Gas chromatograms of samples with other flavour defects are given in Figures 7 to 10.

PTHGC analyses are very useful in the study of aroma compounds which are sufficiently volatile. It should be noted, however, that such methods are utterly inadequate to produce an overview which covers the entire complexity of flavour compounds in milk and milk products (11).

References

1. Morgan, M.E., Day, E.A.: J. Dairy Sci. 48, 1382-1384 (1965).
2. Palo, V., Ilkova, H.: J. Chromatogr. 53, 367-370 (1970).
3. Palo, V.: Chromatographia 4, 55-58 (1971).
4. Palo, V., Hrivnak, J.: Milchwissenschaft 33, 285-287 (1978).
5. Dumont, J.-P., Adda, J.: Ann. Technol. Agric. 27, 501-508 (1978).
6. Marsili, R.T.: J. Chromatographic Sci. 19, 451-456 (1981).
7. Wellnitz-Ruen, W., Reineccius, G.A., Thomas, E.L.: J. Agric. Food Chem. 30, 512-514 (1982).
8. Badings, H.T., Jong, C. de, Wassink, J.G.: HRC & CC 4, 643-646 (1981).
9. Natusch, D.F.S., Thorpe, T.M.: Anal. Chem. 45, 1185A-1194A (1973).
10. Morgan, M.E.: Biotechnol. Bioeng. 18, 953-965 (1976).
11. Badings, H.T., Neeter, R.: Neth. Milk Dairy J. 34, 9-30 (1980).
12. Bassette, R.: J. Milk Food Technol. 39, 10-12 (1976).
13. Allen, C., Parks, O.W.: J. Dairy Sci. 58, 1609-1611 (1975).

SORPTIVE ENRICHMENT AND ANALYSIS OF VOLATILE COMPOUNDS IN DRINKING JUICES

Sieghard Adam

Bundesforschungsanstalt für Ernährung, Institut für Verfahrenstechnik, D-7500 Karlsruhe

Introduction

The collection of headspace vapours and gas extracted samples by sorption on solid porous polymers followed by thermal desorption is a widely used method for gas chromatographic sampling of aroma fractions from foods. Beverages like tea and coffee brews (1-3), beer (4,5), wine (6-8), citrus (9,10) and maracuja juices (11) were analysed for a great number of volatile constituents by using Tenax GC, Porapak Q and Chromosorb 105 as sorbent materials.

The important aroma components are often present at extraordinarily low levels beside a multitude of volatiles at broad concentration ranges in a very complex matrix. Therefore, an ultimate and selective enrichment of the aroma fraction is necessary, in particular when reliable data have to be determined by gas chromatographic / mass spectrometric techniques.

The objectives of this communication have been to present a standardized procedure for analysing volatile constituents in fruit juices and to discuss useful applications in the study of juice technology processes.

Experimental

Samples. The aqueous synthetic test mixture was composed of 3-methyl-1-pentene (1; $0.6\mu l/l$), dimethyl sulfide (2; $0.8\mu l/l$),

420

acetone (3; 0.8μl/l), methyl acetate (4; 0.2μl/l), ethyl acetate (5; 0.3μl/l), methyl alcohol (6; 200μl/l), ethyl alcohol (7;
200μl/l), methyl butyrate (8; 0.6μl/l), 2-methyl-3-buten-2-ol
(9; 0.5μl/l), ethyl butyrate (10; 0.2μl/l), Δ3-carene (11;
0.8μl/l), α-terpinene (12; 0.2μl/l), isopentyl alcohol (13;
0.4μl/l), eucalyptol (14; 0.6μl/l), cis, trans-ocimene (15,16;
1.2μl/l), terpinolene (17; 0.5μl/l), 3-methyl-2-buten-1-ol (18;
0.4μl/l), furfural (19; 12.0μl/l), terpinen-4-ol (20; 5.0μl/l),
α-caryophyllene (21; 0.3μl/l).
The real samples used were: strawberry juices prepared by pressing fruits from different European countries; nectar of black
currants; apple juice; black currant concentrate, and instant
powders of black currant and apple types of commercial origin;
aroma concentrates of different apple juices made available by
a juice manufacturer.

Collection of gas extracted volatiles. Aliquots of the samples
(usually 1 ml) were placed in a gas washing bottle and purged
with helium (99.99 % purity) at 21±1°C for 30 min at a flow
rate of 120 ml/min. The sorption traps, connected to the bottle
outlet by PTFE-connections, consisted of a glass tube (L :
100 mm, O.D.: 6 mm, I.D.: 3 mm) containing 70 mg of freshly
conditioned Porapak Q (80/100 mesh, Waters), Porapak N
(50/80 mesh, Waters), Chromosorb 105 (80/100 mesh, Johns Man-
ville), Tenax GC (60/80 mesh, Enka N.V.) or silica gel
(150/230 mesh, Woelm Pharma). The materials were held in place
by silanized glass-wool plugs.

Collection of headspace vapours. Headspace vapours (volume:
2 ml) above the samples (apple aroma concentrate) were equilibrated at 21±1°C and pushed through a trap containing Tenax GC
(35 mg) by placing a saturated ammonium sulphate solution under
the sample at a flow rate of approx. 1 ml/min (12).

Thermal desorption / cold trapping. The loaded traps were connected to the carrier gas (helium) line of the chromatograph

and purged in the reverse direction for 5 min at a flow rate of
60 ml/min. The gas volumes required for drying the sorbents
(Table 1) were measured by means of a thermal conductivity detector. During the drying period the focusing trap was cooled
to its operating temperature (see Fig. 1).
After drying, the sorption tube was connected to the inlet
transfer line of the chromatograph, heated by a heating coil,
desorbed at $150^{\pm}3^{\circ}C$ and swept into the in-column focusing trap
(13). This trap consisted of a slit steel tube (L = 50 mm,
I.D. = 4 mm) slipped over the first loop of the capillary column and flushed with nitrogen cooled down to $-100^{\pm}5^{\circ}C$ (see
Fig. 1).

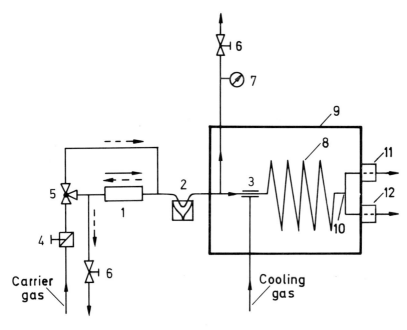

Fig. 1   Scheme of apparatus used for analysing volatile aroma
mixtures. 1: sorbent trap, 2: pre-column cold trap, 3: in-
column cold trap, 4: pressure regulator, 5: three-way valve,
6: needle valve, 7: pressure gauge, 8: separation column,
9: chromatographic oven, 10: effluent split, 11,12: detectors
(FID, ms, human nose).

422

After completion of the desorption process the helium flow was
redirected in order to backflush the sorbent and to supply the
chromatographic column with clean carrier gas. Simultaneously,
the oven was heated to the starting temperature and the condensed sample was evaporated by sucking hot oven air through
the cooling sleeve. The test mixture was desorbed with a helium
volume of 50 ml at a flow rate of 10 ml/min (split ratio = 6:1).
The juices were desorbed with a volume of 90 ml at a flow rate
of 6 ml/min (split ratio = 1:1).
Alternatively, the desorbed material was recondensed in a pre-
column trap consisting of a nickel tube (L = 480 mm, I.D. =
2.4 mm) packed with glass beads and immersed in a Dewar vessel
filled with liquid nitrogen (13). Evaporation was accomplished
by direct resistance heating of the nickel tube (see Fig. 1).

Gas liquid chromatography. Wall-coated open tubular glass columns (L = 50 m, I.D. = 0.3 mm), drawn from soft glass tubes
and statically coated with SP 1000 (Supelco) were used for the
experiments (14). They were operated at 70°C (gas extracted
samples) or 60°C (headspace samples) for 9 min and then tempe-
rature-programmed at a rate of 5°C/min to 170°C where they were
held. The average linear velocity ($\bar{u}_{in}$) of the carrier gas was
20 cm/s at a head pressure of 0.6 bar and 70°C.
After completion of the chromatographic process the sorption
tube was cooled in the helium stream.

Detection. The chromatographed fractions were recorded by a
flame ionization detector (FID). For identification purposes
the outlet of the column was interfaced to a quadrupol mass
spectrometer operated at an ionizing energy of 70 eV (15). An
effluent splitter (Labormechanik Gerstel) was connected to the
column outlet to allow simultaneous FI-detection and odour
description by sniffing the chromatographed vapours.

Results and Discussion

Testing of sorbent materials. An aqueous synthetic mixture of
21 volatile substances approximating the composition of black
currant mash, was used for studying the sorption / desorption
properties of various sorbents. The upper chromatogram in
Fig. 2 was obtained by using gas extraction and sorption on
Porapak Q, thermal desorption and cryogenic band focusing with
an in-column trap. An identical chromatogram was obtained when
Chromosorb 105 was utilised as the sorbent. Retention indices
and mass spectra of the numbered peaks were in accordance with
those of reference compounds sampled from organic solvents. The
unmarked small peaks in the chromatogram obviously originate
from impurities of the supplied chemicals. The standard error
for repeated analyses of compounds 9-21 was less than 10 %
each. Correlation coefficients greater than 0.98 were found
between the detector response and the content of compounds
9-21 in the starting mixture.
The use of Tenax GC in place of Porapak Q revealed incomplete
recovery of 2-methyl-3-buten-2-ol (peak 9) which is only partly
retained by the former.
In an effort to recover also the low-boiling components 1 and
2, 3-methyl-1-pentene and dimethyl sulfide, and to enhance the
recovery of the substance 3 - 8 other sorbents were examined.
The use of Porapak N which distinguishes by a strong affinity
for low-molecular weight volatiles led to a congruent peak pattern, when the extended drying period for the loaded sorbent
was taken into account (Table 1). Omission of the purge-drying
cycle and desorption of the humid test sample led to the central chromatogram in Fig. 2. The chromatogram revealed peak
shifts (16) for many compounds, most of them towards decreased
retention times. Although partial recovery of methanol and
ethanol was achieved, the desired retention of the remaining
low-boiling volatiles was not obtained.
Silica gel which tends to fix water at room temperature proved
to be insufficient as the sorption / desorption material.

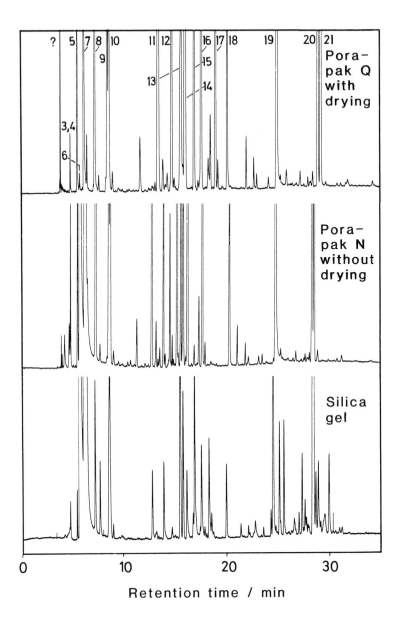

Fig. 2   Gas chromatograms of volatile components from an aqueous test mixture, collected on various sorbents and subsequently thermally desorbed. Assignment of numbered peaks: see "Experimental".

Table 1   Volumes of helium required for drying various
sorbents (70 mg) after sorption of aqueous vapours at 21°C

Sorbent	Purge-drying volume (l)
Tenax GC	0.3
Porapak Q	0.3
Chromosorb 105	0.6
Porapak N	1.5
Silica gel	2.0

Irreversible effects during collection, heating and chromatography are obviously due to the high water content leading to
the poor quality of the corresponding chromatogram (Fig. 2).

Fingerprint analyses. The fruit processing industry has made
great efforts to control the quality of raw materials. To
jugde the quality of the juices sensory analyses are the
method of choice which usually yield reliable results but in-
volve considerable expense. Therefore, attempts are being made
to supplement, or possibly substitute, sensory jugdement by
objective measuring techniques.
Fig. 3 shows chromatograms for juices produced from strawberries of different origins. Current studies are being under-
taken to correlate the obtained peak patterns (fingerprints)
with sensory assessments.
Fig. 4 demonstrates the differences of headspace chromatograms
produced from aroma concentrates of juices from apples of different origins. Sample A displaying the highest peak intensities in the chromatogram was described as the most typical one
by subjective odour analysis. Samples B and C which contain
lower amounts of volatile compounds were of minor quality.

Study of concentration processes. During the past two decades
increasing quantities of fruits have been consumed as processed

426

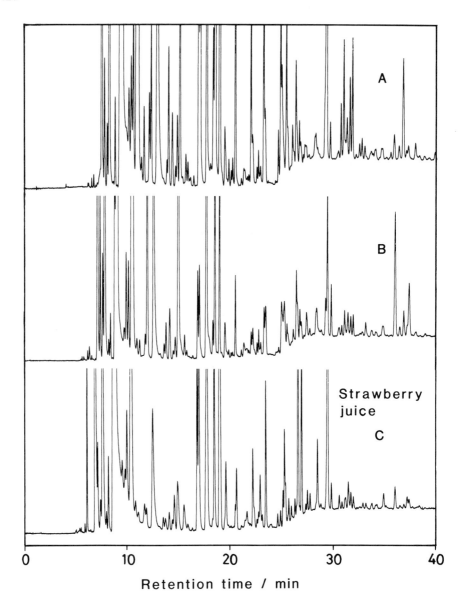

Fig. 3   Gas chromatograms of volatile constituents in fresh juices from strawberries of different origins (A, B, C).

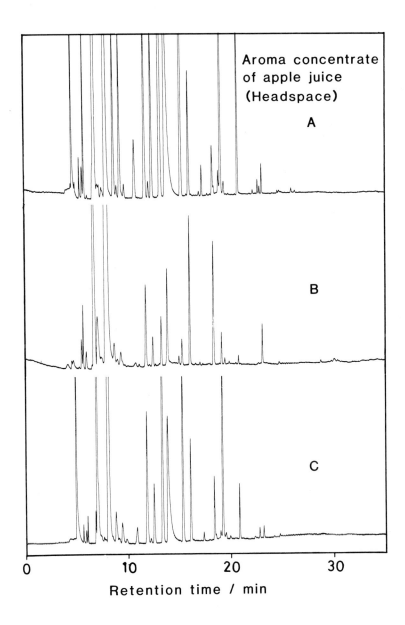

Fig. 4   Gas chromatograms of headspace vapours over aroma
concentrates of juices from apples of different origins (A,B,C).

products. The shift from fresh fruit to processed products has
been due to the all-year availability of the latter and their
improved uniform quality (17). The removal of water from press-
juices by evaporation ist the most economical and most widely
used method of concentration. During this operation, however,
most of the volatile aroma compounds are removed together with
the water vapour yielding products of low sensory value. The
losses can be partly compensated for by "cut-back" of fresh
juice to the concentrate (18). The seperation of pulp and serum
followed by the concentration of the serum fraction and recom-
bination of pulp and concentrated serum is another process
which partly restores the quality of the original juice (18).
High degrees of aroma retention are obtained by dividing the
starting material into the juice concentrate and the water /
aroma fraction which is subsequently further enriched by recti-
fication. The aroma concentrate obtained is finally added back
to the juice concentrate (18).

Fig. 5 shows chromatograms of different products derived from
black currants. The upper picture represents a typical distribution of volatiles in a commercial drinking juice (nectar).
Odour description by sniffing the vapours emerging from the end
of the chromatographic column (see Fig. 1) revealed that at a
retention time of 12 min one (or more) component (s) is (are)
responsible for the typical aroma of juices from black currants.
Comparison with the peak profile from a commercial juice concentrate of black currants, diluted to the ready-to-drink concentration, shows a low amount of fast-chromatographing components in the concentrate. These volatiles including the compounds of characteristic odour are obviously lost during the
concentration process leading to a final product of low sensory
quality.

Study of drying processes. From the viewpoints of increasing
product stability and decreasing transport and storage costs it
is advantageous to manufacture dried products from juice concentrates. Hitherto dewatering processes have always been

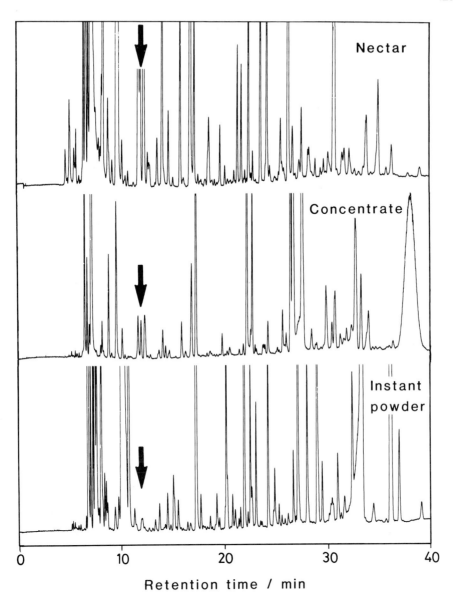

Fig. 5 Gas chromatograms of volatile constituents in black currant drinking juices (nectar, diluted concentrate and dispersed instant powder). The arrows denote the chromatographic region of the characteristic odour of black currant juice.

430

accompanied by effects influencing adversely the product quality (19). Insufficient aroma retention may indeed be considered
as the critical point in the drying technology of the majority
of foods.
Freeze-drying is characterized by relatively high aroma retention (20). The drying rate, however, is too low for practical
purposes.
Slush-drying is an interesting method providing high drying
rates combined with satisfactory retention of volatile aroma
compounds (21). Spray-drying constitutes the most widely used
process mainly applied to produce fruit powders.

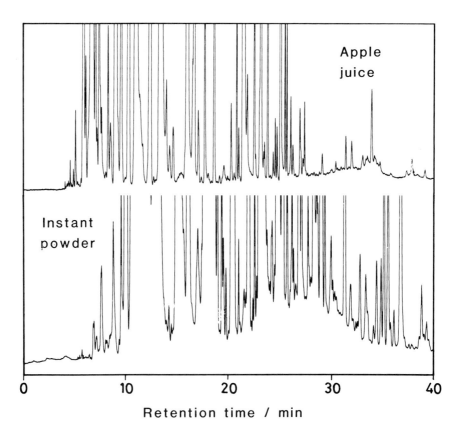

Fig. 6   Gas chromatograms of volatile constituents in an
apple juice and a dispersed instant powder ("apple type").

A further improvement of drying techniques seems very promising provided that refined analytical methods are used to compare results and to optimize the process parameters accordingly.

Fig. 6 shows a comparison between a commercial apple juice and a commercial instant apple powder dispersed in water. The powder contains less volatiles at a retention time less than 10 minutes and larger quantities of volatiles after long retentions. Preliminary studies revealed that synthetic aroma compounds had been added to the instant powder during production.
The chromatogram of an instant powder "black currant type" is also shown in Fig. 5 (lower chromatogram). This powder obviously contains some volatile components which are also present in natural juices made from black currants. The typical natural aroma compounds, however, are absent.

References

1. Vitzthum, O.G., Werkhoff, P.: G. Charalambous (Ed.), Proc. Analysis of Foods and Beverages, Headspace Techniques, Chicago, Aug.29-Sept.2, 1977, Academic Press, New York, San Francisco, London, 1978, pp.115-133.

2. Tassan, C.G., Russell, G.F.: J. Food Sci. 39, 64-68 (1974).

3. Ott, U., Liardon, R.: P. Schreier (Ed.), Flavour 1981; Proc. Internat. Conference/3rd Weurman Symposium, Munich, April 28-30, 1981, Walter de Gruyter, Berlin, New York, 1981, pp.323-338.

4. Jennings, W.G., Wohleb, R., Lewis, M.J.: J. Food Sci. 37, 69-71 (1972).

5. Leppänen, O., Denslow, J., Koivisto, T., Ronkainen, P.: P. Schreier (Ed.), Flavour 1981; Proc. Internat. Conference 3rd Weurman Symposium, Munich, April 28-30, 1981, Walter de Gruyter, Berlin, New York, 1981, pp.361-368.

6. Bertuccioli, M., Montedoro, G.: J. Sci. Food. Agric. 25, 675-687 (1974).

7. Williams, P.J., Strauss, C.R.: J. Inst. Brew. 83, 213-219 (1977).

8. Simpson, R.F.: Chromatographia 12, 733-736 (1979).

432

9. Schultz, T.H., Flath, R.A., Mon, T.R.: J. Agr. Food Chem. 19, 1060-1065 (1971).

10. Lund, E.D., Dinsmore, H.L.: G. Charalambous (Ed.), Proc. Analysis of Foods and Beverages; Headspace Techniques, Chicago, Aug.29-Sept.2, 1977, Academic Press, New York, San Francisco, London, 1978, pp.135-185.

11. Murray, K.E.: J. Chromatogr. 135, 49-60 (1977).

12. Von Sydow, E., Andersson, J., Anjou, K., Karlsson, G., Land, D., Griffiths, N.: Lebensm. - Wiss. u. Technol. 3, 11-17 (1970).

13. Adam, S.: J. High Res. Chromatogr. 6, 36-37 (1983).

14. Jennings, W.G., Yabumoto, K., Wohleb, R.H.: J. Chromatogr. Sci. 12, 344-348 (1974).

15. Koller, W.D., Tressl, G.: J. High Res. Chromatogr. 3, 359-360 (1980).

16. Burns, W.F., Tingey, D.T., Evans, R.C.: J. High Res. Chromatogr. 5, 504-505 (1982).

17. Deshpande, S.S., Bolin, H.R., Salunkhe, D.K.: Food Technol. 36 (5), 68-82 (1982).

18. Pala, M., Bielig, H.J.: Industrielle Konzentrierung und Aromagewinnung von flüssigen Lebensmitteln, Universitätsbibliothek T.U. Berlin, Abteilung Publikationen, Berlin 1978.

19. Thijssen, H.A.C.: Lebensm.-Wiss.u.Technol. 12, 308-317 (1979).

20. Flink, J.M.: J. Agric. Food Chem. 23, 1019-1026 (1975).

21. Chandrasekaran, S.K., King, C.J.: Chem. Engin. Progr. Symp. Ser. 67, 122-130 (1971).

VOLATILES IN RELATION TO AROMA IN THE BERRIES OF RUBUS ARCTICUS COLL.

Heikki Kallio and Anja Lapveteläinen

Department of Chemistry and Biochemistry, Laboratory of Food Chemistry, University of Turku, SF-20500 Turku 50, Finland

Timo Hirvi, Meri Suihko, Erkki Honkanen

Technical Research Centre of Finland, Food Research Laboratory, Biologinkuja 1, SF-02150 Espoo 15, Finland

Introduction

Aroma of the berries of arctic bramble, Rubus arcticus coll. has been investigated quite intensively. The best strains among these studied belong to R. arcticus L. subsp. arcticus (1-3). However, the fully ripe berries of subsp. stellatus (Sm.) Boiv. are big, red and juicy, but have weaker aroma than the berries of the arcticus subspecies (1,4). In order to develop a new aromatic, hardy berry with good yield Larsson has crossed the two subspecies with each other several times after 1952 (4,5). The work has been continued since the spring 1969 also at the Agricultural Research Centre of Finland (6). The composition of the aroma of the hybrides has also been studied (7). Both in the quantity and the quality of the aroma, the hybrides were between the parents.

When analysing the aroma of small samples of berries of different strains or clones of one species, a rapid and reliable method is needed. The chemical data analysed should also correlate to the results of the sensory analyses. If the correlation is good, the sensory tests can preliminarily be replaced by the chemical analyses, especially in shortage of berries. Hirvi (8) has studied the quality of strawberry cultivars and Hirvi and Honkanen (9) that of blueberry cultivars by carrying out the analyses of volatiles with the mass spectral SIM (Selected Ion Monitoring) technique, and by comparing the results to the sensory qualities. Knowledge of this correlation is needed especially when producing new hybrides of fruits and berries for commercial use.

The aim of this study was to compare the sensory quality and chemical composition (GLC-MS, SIM-technique) of the volatiles in the berries of arctic bramble, Rubus arcticus coll., and to find out the natural variation of the parametres among the various strains of the subsp. arcticus. Also a comparison to subsp. stellatus and to a hybrid between the subspecies was made.

Experimental

1) Berries. The berries (crop 1982) of eight different strains of arctic bramble, Rubus arcticus coll. were studied. The cultivars 'Mespi' (10) and 'Pima' (11) of subsp. arcticus were cultivated at Hiskilä farm (Leppävirta, Finland, 62° 36´N, 27° 33´E) and at the Agricultural Research Centre, South Savo Experiment Station (Mikkeli, Finland, 61° 40´N, 27° 14´E). All the following berries were grown in Mikkeli using normal cultivating practice. The two other strains of subsp. arcticus investigated were a cultivar 'Mesma' (10) and a wild strain (ST_2) originating from Maaninka (11). The berries of a clone of subsp. stellatus (6) and of a hybrid subsp. stellatus x subsp. arcticus, code 086 (12) were also studied. The berries were picked optimally ripe, frozen and stored at -25^OC.

2) Isolation of volatiles. The berries were allowed to thaw at $+5^O$C for 16 hours and one hundred grams of the berries were pressed for each replicate analysis (yield 62±5 %, w/w). 30 g of the juice was extracted three times with a mixture of pentane-diethylether (40 ml, 1:2, v/v) for three minutes in a separating funnel. The extracts were combined, dehydrated with Na_2SO_4 and concentrated with a Vigreaux column to 3 ml followed by concentration in nitrogen stream to 0.5 ml. The

concentrates were stored under nitrogen at 5°C.

3) <u>Quantitation of aroma compounds</u>. For the analyses we selected 11
   volatiles, which have been previously shown to be important for the
   aroma of arctic bramble (2, 13). The quantitation was performed using
   a combined GLC-MS (Hewlett-Packard 5992) with the mass fragmento-
   graphic selected ion monitoring (SIM) technique. The compounds analysed
   and ions monitored are shown in Table 1.

   A fused silica capillary column coated with OV-351 phase (length 25 m,
   I.D. 0.3 mm) and on-column injection technique were used. The condi-
   tions were: Linear temperature programming from 50°C to 230°C,
   10°C·min^{-1}, flow rate of the carrier He was 2 ml·min^{-1}, electron

compound	selected ion (m/e)	intensity in mass spectrum (%)
2,5-dimethyl-4-methoxy-3(2H)-furanone	142	35
2,5-dimethyl-4-hydroxy-3(2H)-furanone	128	95
3,7-dimethyl-1,6-octadien-3-ol (linalool)	71	100
2-heptanol	45	100
3-methyl-2-buten-1-ol	71	100
cis-3-hexen-1-ol	82	40
3-methylbutanoic acid	60	100
2-phenylethanol	91	100
4-allyl-2-methoxyphenol (eugenol)	164	100
4-hydroxy-3-methoxybenzaldehyde (vanillin)	152	100
3-methyl-2-butenyl acetate	68	100

Table 1. Typical aroma compounds analyzed in the berries of arctic
bramble, the ions monitored and their relative intensities
in the mass spectra (70 eV).

436

current 300 μA, ionization energy 70 eV and sweep time 100 μs·ion^{-1}.
The quantification was performed by comparing the signals of the
SIM-peaks to those obtained within the mixture of the reference com-
pounds.

4) Reference compounds. 3-methyl-2-butenyl acetate, 2,5-dimethyl-4-
methoxy-3(2H)-furanone and 2,5-dimethyl-4-hydroxy-3(2H)-furanone were
synthesized and their purity was checked with GLC. The other reference
compounds were commercial products.

5) Sensory analyses. For the sensory analyses the juice was separated
as for the chemical analyses and diluted with tap water (1:2). The
trained panel composed of ten persons familiar with the flavour of
arctic bramble. Intensity, character, overall impression of odour and
off-odours (all eight samples) as well as sweetness, sourness, off-
tastes and overall impression of taste (all six subsp. arcticus samples)
were evaluated using a graphical scaling method (a 100 mm long scale)
of Stone et al. (14). In each session four samples (10 ml, 20°C) were
presented in randomized order including the variety 'Pima' (cultivated
in Mikkeli) as a hidden reference. The evaluations were carried out
under red light. The samples were served in odourless 50 ml beakers
covered with a glass plate. Correlations between the sensory and
instrumental results were studied by correlograms and regression
analysis.

Results and discussion

1) Reliability of the chemical analyses. Table 2. presents the standard
deviations of the chemical analyses. Four replicates of one lot of
berries (wild strain ST$_2$ of subsp. arcticus) were analysed separately.
Also five parallel analyses of one aroma concentrate were performed
to obtain the reproducibility of the method. The results of the
replicate samples differed considerably from each other. The standard
deviation was high (11-42 %) in most cases. One reason for this is

compound	content in the juice, mg·kg^{-1}					
	four berry samples, analyzed separately			five replicate GLC-MS-analyses		
	$\overline{X}$	S.D.	S.D./$\overline{X}$ %	$\overline{X}$	S.D.	S.D./$\overline{X}$ %
2,5-dimethyl-4-meth-oxy-3(2H)-furanone	3.1	0.70	23	2.3	0.15	7.1
2,5-dimethyl-4-hydr-oxy-3(2H)-furanone	10	1.8	18	-	-	-
linalool	0.70	0.25	34	0.60	0.06	11
2-heptanol	0.65	0.20	33	-	-	-
3-methyl-2-buten-1-ol	0.80	0.35	42	1.1	0.04	3.5
cis-3-hexen-1-ol	0.04	0.02	35	-	-	-
3-methylbutanoic acid	1.5	0.15	12	1.6	0.15	8.0
2-phenylethanol	0.10	0.02	23	-	-	-
eugenol	0.01	0.00	29	0.01	0.00	0.00
vanillin	0.05	0.01	20	0.04	0.00	12
3-methyl-2-butenyl acetate	0.45	0.05	11	-	-	-

Table 2. Reliability of the analyses (wild strain subsp. arcticus ST$_2$ grown in Mikkeli)

the different stage of maturity of the berries in different lots. The concentrations of many important aroma compounds in arctic bramble are changing even at the optimal ripeness (3). Thus, special care in the sampling and critical evaluation before conclusions is necessary.

2) Contents of aroma compounds. Figure 1. shows an example of the SIM analyses: 1a, a standard mixture of 3-methyl-butanoic acid, eugenol, vanillin, 3-methyl-2-buten-1-ol, linalool and 2,5-dimethyl-4-methoxy-3(2H)-furanone (each 10 mg·kg^{-1} in the standard solution); 1b, the corresponding arctic bramble compounds; 1c, a standard mixture of 3-methyl-2-butenyl acetate, 2-phenylethanol, cis-3-hexen-1-ol, 2-heptanol and 2,5-dimethyl-4-hydroxy-3(2H)-furanone; 1d, the corresponding arctic bramble compounds. The SIM-tecnique is suitable for comparative aroma analyses. It is fast and reliable in qualitative

438

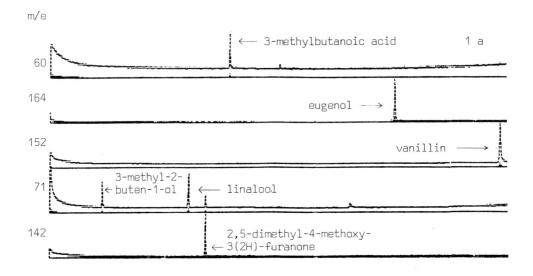

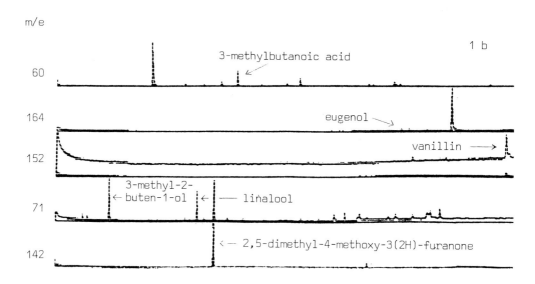

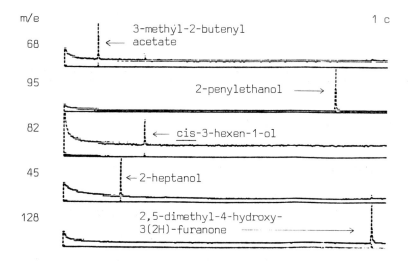

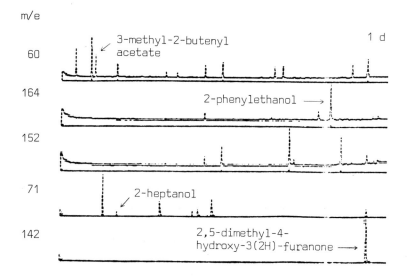

Figure 1.   Examples of SIM-analyses. Figures 1a and c
represent mixtures of reference compounds and
1b and d of aroma compounds in arctic bramble
(wild strain ST_2).

as well as in quantitative research. Selection of the monitored ion
or, if needed, ions, is naturally the point of possible mistake.

The contents of the eleven compounds, typical to arctic bramble (2,3),
analysed in the eight varieties and hybrides, are presented in table 3.
2,5-dimethyl-4-methoxy-3(2H)-furanone is known to be one of the main
aroma compounds in arctic bramble (15), as was also observed in this
work. The content of 2,5-dimethyl-4-hydroxy-3(2H)-furanone
(5-14 $mg \cdot kg^{-1}$) was also very high and in some varieties it even
exceeded the content of the former compound. The odour threshold
values for both of these components are low: for 2,5-dimethyl-4-
methoxy-3(2H)-furanone 0.01 $mg \cdot l^{-1}$ (16) and for 2,5-dimethyl-4-hydroxy-
3(2H)-furanone 0.00003 $mg \cdot l^{-1}$ (17). It is therefore evident that
both furanones are involved in the formation of the strong aroma of
arctic bramble, because they formed 69 to 98 per cent of the total
contents of the eleven compounds analysed.

Hirvi et al. (18) noticed these furanones to be labile at room temperature. Thus the isolation and storage conditions have to be always carefully controlled.

3) Sensory analyses of odour and taste. Table 4, presents the results of
the odour and taste evaluations as the means of ten analyses. The
digits correlate to the overall length of the graphical scale (100 units
long). The juice had to be diluted because the strong odour prevented
the analysis and evaluation. The results can be regarded as reliable
because those of the reference variety 'Pima' were very similar in
each experiment.

According to the character and the intensity of the odour the cultivar
'Mesma' was regarded as the best variety. The overall impression was
the second in order after the berries of 'Mespi' grown in Mikkeli.
The sourness of 'Mesma' lowered its overall impression of taste,
however, to the next worst among all the berries of subsp. arcticus.
The intensity of odour was among all strains of this subspecies at
quite the same level (on the scale between 58 and 70). The corre-

content in the juice, mg·kg^{-1}

compound	subsp. arcticus						subsp. st × subsp. a, 086[b]	subsp. stellatus[b]
	'Pima,[a]	'Mespi,[a]	'Pima,[b]	'Mespi,[b]	'Mesma,[b]	wild ST$_2$[b]		
2,5-dimethyl-4-methoxy-3(2H)-furanone	12	20	3,8	11	8.1	3.1	2.8	13
2,5-dimethyl-4-hydroxy-3(2H)-furanone	6.3	11	4.9	8.1	14	10	4.8	0.75
linalool	0.30	0.65	0.07	0.45	0.25	0.70	0.08	0.03
2-heptanol	2.7	0.90	1.6	1.0	3.4	0.65	1.9	–
3-methyl-2-buten-1-ol	0.95	0.95	0.07	0.40	1.5	0.80	0.06	0.02
cis-3-hexen-1-ol	0.15	0.25	0.10	0.30	0.30	0.04	0.06	–
3-methylbutanoic acid	0.14	0.15	0.03	0.10	0.20	1.5	1.1	0.30
2-phenylethanol	0.75	0.40	0.80	0.55	0.55	0.10	0.10	+
eugenol	+	+	+	+	+	0.01	0.01	+
vanillin	0.02	0.02	0.01	0.03	0.02	0.05	0.03	0.04
3-methyl-2-butenyl-acetate	0.15	0.20	0.02	0.05	0.20	0.45	–	–

Table 3. Contents of eleven aroma compounds in the juice of arctic bramble (means of one to four analyses)

[a] grown in Leppävirta, [b] grown in Mikkeli

442

strain of arctic bramble	number of the strain	odour estimation				taste estimation			
		intensity	character	off-odour	overall impression	sweetness	sourness	off-taste	overall impression
'Pima'[b]	1	59	60	10	61				
		60	56	9	61	35	70	10	50
		61	64	7	64	26	58	9	48
		62	64	9	62	30	63	8	52
	$\bar{X}$	61	61	9	62	30	64	9	50
	S.D.	1.3	3.8	1.3	1.4	4.5	6.0	1.0	2.0
'Pima'[a]	2	61	66	9	70	56	41	4	78
'Mespi'[a]	3	65	66	4	64	55	40	4	78
'Mespi'[b]	4	58	76	5	82	51	48	3	74
'Mesma'[b]	5	70	76	4	77	25	78	8	56
wild ST_2[b]	6	61	64	6	71	34	65	4	70
subsp. st × subsp. a, 086[b]	7	61	51	21	48	-	-	-	-
subsp. stellatus[b]	8	49	43	28	44	-	-	-	-

Table 4. Sensory analysis of odour and taste of the juice of arctic bramble

[a]grown in Leppävirta, [b]grown in Mikkeli

sponding value of subsp. stellatus was clearly lower, 49. The odour of the hybrid 086 was evaluated as strong as that of 'Pima' cultivars and the wild strain subsp. arcticus, ST_2. In all, the berries of subsp. stellatus and also of the studied hybrid were less delicious than the berries of the ectual arctic brambles, and had also a trace of an off-odour not typical to arctic bramble.

The overall impression of the taste was best in the 'Pima' and 'Mespi' cultivars grown in Leppävirta. They were also sweeter and less sour than the other berries studied. As the 'Pima' from Mikkeli was evaluated as the least delicious (both odour and taste) it is very clear, that the growth conditions and micro climate are of great importance to the sensory quality of arctic bramble.

4) Correlation between the sensory and chemical parametres.
The correlations between the total amount of the eleven compounds and
each item of the sensory data were not good the coefficients being:
to overall impression r = 0.556, to intensity r = 0.564 and to
character of odour r = 0.649. The correlation between the content of
2,5-dimethyl-4-hydroxy-3(2H)-furanone and the odour is presented in
Figure 2. The content of this furanone compounds describes well the
intensity of the odour (r = 0.867, p = 0.003), somewhat less the
character of the odour (r = 0.821, p = 0.006) and the overall impression of the odour (r = 0.751, p = 0.016).

Contrary to the previous results (2) we could not find a good
correlation between the content of 2,5-dimethyl-4-methoxy-3(2H)-
furanone and the odour evaluations. However, some dependence of

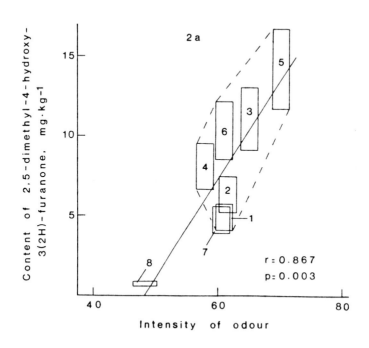

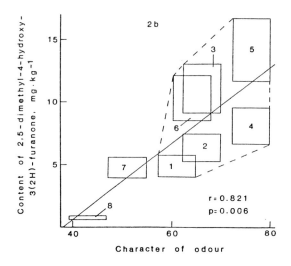

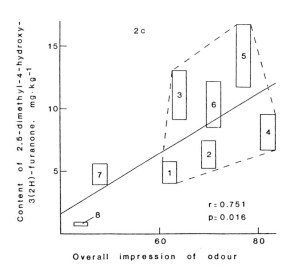

Figure 2. Correlation of the content of 2,5-dimethyl-4-hydroxy-3(2H)-furanone to the odour of arctic bramble. Each quadrangle represents one strain of the berry. The horizontal sides are the doubled standard deviations of the corresponding sensory properties (Table 4.) and the vertical sides that of the corresponding chemical factor, in percentages (Table 2). The dotted area shows the typical variation of R arcticus subsp. arcticus.

445

the overall impression of the taste on the contents of the above mentioned methoxy furanone compound seems evident (Fig. 3). In the earlier works the aroma compounds were isolated by steam-destillation, which does not give a complete picture of the aroma (2,3,15).

In the work the correlations between the contents of the aroma compounds and the sensory evaluations have been assumed to be linear. This hypothesis may be inaccurate and incorrect. However, Figures 2 and 3 give us a useful and generally acceptable basis for the understanding of the natural variation among the berries of arctic bramble from different origins. The most important point is of course the selection of the compounds to be examined; the calculated aroma values are not necessarily the only right basis. In an earlier work Hirvi (8) noticed more significant correlations, than in this work, while studying some strawberry varieties using the same methods.

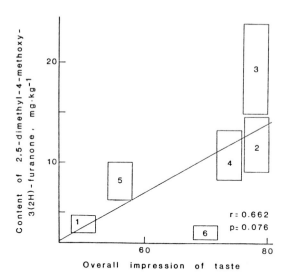

Figure 3.   Correlation of the content of 2,5-dimethyl-4-methoxy-3(2H)-furanone to the overall impression of taste of the berries of subsp. arcticus. For explanations, see Fig. 2.

446

Acknowledgements

The authors express the best thanks to mrs. Pirjo Dalman, M.Sc.
for donating us the berries. Also Mr. Timo Alanko, M.Sc. is acknowledged
for the statistical analyses.

References

1. Kallio, H.: Rep. Kevo Subarctic Res. Stat. 12, 60-65 (1975).
2. Kallio, H.: J. Food Sci. 41, 555-562 (1976).
3. Kallio, H.: J. Food Sci. 41, 563-566 (1976).
4. Larsson, E.G.K.: Hereditas 63, 283-351 (1969).
5. Larsson, E.G.K.: Bot. Notiser 133, 227-228 (1980).
6. Hiirsalmi, H.: Puutarhantutk. Lait. Tied. 22, 31-47 (1979).
7. Kallio, H., Laine, M., Huopalahti, R.: Rep. Kevo Subarctic Res. Stat.
    16, 17-22 (1980).
8. Hirvi, T.: Lebensm.-Wiss. u. -Technol. 16, 157-161 (1983).
9. Hirvi, T., Honkanen, E.: Z. Lebensm. Unters. Forsch. 176,
    346-349 (1983).
10. Ryynänen, A.: Ann. Agric. Fenn. 11, 170-173 (1972).
11. Ryynänen, A., Dalman, P.: Ann. Agric. Fenn. in press (1983).
12. Hiirsalmi, H., Säkö, J.: Acta Hort. 112, 103-108 (1980).
13. Pyysalo, T., Suihko, M., Honkanen, E.: Lebensm.-Wiss. u.
    -Technol. 10, 36-39 (1977).
14. Stone, H., Sidel, J., Oliver, S., Woolsey, A., Singleton, R.C.:
    Food Technol. 28, 24-34 (1974).
15. Kallio, H., Honkanen, E.: Proc. IV Int. Congress Food Sci. and
    Technol. 1, 84-92 (1974).
16. Pyysalo, T., Honkanen, E., Hirvi, T.: J. Agr. Food Chem. 1, 19-22
    (1979).
17. Honkanen, E., Pyysalo, T., Hirvi, T.: Z. Lebensm. Unters. Forsch. 171,
    180-182 (1980).
18. Hirvi, T., Honkanen, E., Pyysalo, T.: Lebensm.-Wiss. u. -Technol. 13,
    324-325 (1980).

ANALYSIS OF COFFEE HEADSPACE PROFILES BY MULTIVARIATE STATISTICS

Rémy Liardon, Ursula Ott, Nicole Daget
Nestlé Research Department
CH-1814 La Tour-de-Peilz, Switzerland

Introduction

The volatile fraction of a food usually consists of a complex mixture of components which have been formed at the various stages of the product life. Therefore, many characteristics of a food are reflected in its volatiles. However, most often the complexity of the volatile fraction is such that the information hidden therein can only be retrieved and interpreted with the help of multivariate statistics.

Over the years, many attempts have been made to make use of food volatile profiles. In one of the earlier studies, coffee samples with distinct flavour characteristics were discriminated by means of GC peak ratios selected by stepwise discriminant analysis (1). Later on, similar investigations were performed on many different products and a wide range of parameters were tentativley related to the composition of the volatile fraction (2-12). More recently, the evolution of the analytical techniques has contributed to a renewed interest in this kind of studies. This includes the development of new sampling procedures like headspace trapping and the generalization of high performance GC capillary columns. The application of statistical treatments to the data obtained with these improved techniques would more likely provide meaningful results than was the case in the first studies.

Since the pioneer work of Powers and Keith (1) and Biggers et al. (13), no further attempt had been made to investigate coffee volatiles using the statistical approach. This prompted us to undertake a new study based on updated techniques. First, we developed the instrumentation and methodology for the routine analysis of coffee headspace by capillary GC (14). In the following stage, several experiments were carried out in which series of coffee headspace profiles were collected and submitted to various multivariate statistical treatments. Different possible

Analysis of Volatiles
© 1984 Walter de Gruyter & Co., Berlin · New York – Printed in Germany

applications have been considered and are presented here. They include the discrimination of coffee samples, investigations on the influence of different parameters on coffee volatiles, and the correlation of sensory and chemical data.

Methodology

Sampling and analysis

The procedure and the instrumentation developed for coffee headspace analysis have been described elsewhere (14). Basically, sampling consisted in collecting the headspace components of freshly brewed coffee in adsorbing traps packed with Tenax-GC. The collected volatiles were analysed on a modified HP 5880 gas chromatograph equipped with a 100 m x 0.25 mm i.d. glass capillary column coated with UCON 50-HB-5100. The thermal desorption of the components and the subsequent GC run were performed under the control of HP 5880 microprocessor which ensured an optimum reproducibility of the chromatographic profiles. Several headspace samples were also run under the same chromatographic conditions on a HP 5992 GC/MS. A majority of the components could be identified on the basis of their mass spectra.

Data processing

The gas chromatograph was connected to a HP 3354 Laboratory Automation System which performed the acquisition and integration of the GC data in real-time. The resulting reduced data were transferred onto a HP 3000 computer by means of small tape cartridges. At this stage, an automatic peak recognition procedure was applied to the entire set of chromatograms belonging to the same experiment. In this process, each GC peak was assigned an identification number valid in all considered chromatograms. Then the entire data set was reformated and stored in a BMDP compatible file (15). Only the absolute surface areas, as determined by the integration routine, were retained as pertinent data. The normalization of peak intensities in percent of the total area was not considered since it would have suppressed all information on the actual concentration of the components in the various coffees.

Statistical analysis

Various multivariate statistical procedures were used in this study, namely the stepwise discriminant analysis (SDA), canonical analysis (CA), stepwise regression analysis (SRA), as well as the Procrustes analysis (16). Part of the treatments were made with routines belonging to the BMDP package (15). Other programs which had been developed in our group were also used, in particular for CA and Procrustes analysis. The particularity of these two routines was to provide a vectorial representation of the original variables in the canonical space (17).

Most of those programs could not accommodate more than 50 to 70 variables. In order to meet this requirement, only the components appearing in all chromato-grams of the experimental set were retained for the statistical treatment.

Discrimination of Coffee Variety and Roasting

The area where the application of multivariate statistics to food chemical profiles has been most successful is the discrimination of product categories among samples of the same commodity. Based on this experience, our first study was designed to evaluate the possibility of using headspace profiles to discriminate coffee samples differing by variety or roasting level. The detailed results of this study have been presented elsewhere (18). They are outlined here.

Fifteen coffees were selected for this experiment. They consisted of three varieties of beans which were roasted at five predetermined levels ranging from light to dark roast. The three varieties were: Kenya (wet processed Arabica), Santos (dry processed Arabica, Brazil), and Ivory Coast (dry processed Robusta). Three to five headspace profiles were recorded for each coffee, resulting in a total of 55 chromatograms. This data set was processed as described in the preceding section. After removal of the minor peaks, which could not be found in all chromatograms, the reduced data set still comprised 66 headspace components.

The statistical treatment performed on these data followed a classical approach: first SDA was applied to select the components which would significantly contribute to discriminate the 15 coffees within the 55 profiles, then the selected data subset

450

was analysed by CA. This permitted to evaluate its discriminating power in terms of the probabilities associated with the distances between any two coffees in the canonical space (Mahalanobis distances). Another test for the discriminating power consisted in a classification attempt performed by the BMDP7M program. Using the information provided by the selected components, this program would tentatively assign each experimental profile to one of the 15 coffees. The success of this classification was measured by the percentage of correctly assigned profiles.

The complete procedure was repeated several times while forcing the selection of different components at the first step of SDA. The different subsets obtained in this way comprised 18 to 24 components. Fig. 1 shows the canonical configuration of one of these subsets. This is a projection of the canonical space on the plane spanned by the first two canonical axes. Together these two axes carried 65% of the total discrimination. In this figure, dots indicate the average position of each coffee. Also shown are the vectors representing the selected components. In that example, all the probabilities associated with the Mahalanobis distances were found to be far below .01, which indicated that all 15 coffees were completely discriminated. Confirming this result, 98% of the profiles were correctly assigned in the classification attempt.

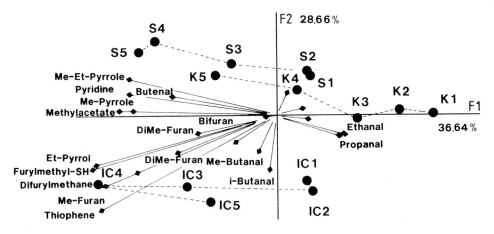

Fig. 1. Canonical analysis of 21 coffee headspace components selected by SDA. Coffee dots are labelled according to variety and roasting: K = Kenya, S = Santos, IC = Ivory Coast; increasing roasting levels are numbered from 1 to 5. The percentage value reported for each axis is the portion of total discrimination carried by that axis.

As it turned out, very similar results were achieved with the different subsets selected at that level. Further attempts were made to determine the influence of the size of the subset on the discriminating power. Subsets containing 9 and 6 components were successively analysed by CA. In both cases the level of discrimination was excellent (Mahalanobis probabilities below .01). The success rate for the classification was 91% with 9 components and 86% with 6 components.

As a conclusion, this experiment confirmed that coffee parameters like variety and/or roasting could be discriminated by means of headspace profiles. Moreover, this result could be achieved on the basis of the combined information provided by a fairly small number of components.

Influence of Product Parameters on Coffee Volatiles

Another interesting aspect of the CA treatment was to show in a single analysis the dependence of the various components on the product parameters (e.g. roasting or variety). This information was derived from the superposition of the product configuration and the vectorial representation of the components in the canonical space. This can be illustrated on the basis of Fig. 1.

In this projection, the trajectories followed by each coffee variety clearly designated the first canonical axis as the roasting axis. Conversely, the second canonical axis, or more precisely the plane orthogonal to the first axis, represented the elements discriminating the three varieties. The orientation of each component with respect to these two main directions would reflect the extent of its dependence on the roasting or variety parameter. As a matter of fact, all the observations which could be made on the basis of this purely statistical analysis were in excellent agreement with otherwise available data on coffee chemistry (18). For example, it was noticed that the vectors corresponding to chemically related components usually were grouped in tight bundles, as could be expected due to their common origin or formation pathways. On the other hand, aliphatic aldehydes appeared to be split in two totally unrelated groups. As it turned out, this configuration was consistent with the existence of two distinct formation mechanisms. In coffee ethanal and propanal originate from sugar pyrolysis, while isobutanal, 2-methyl and

3-methylbutanal are the Strecker degradation products of valine, isoleucine and leucine. Interestingly, the orientation of the branched aldehydes in the canonical space seemed to reflect differences in the amino acid composition of the two botanical varieties of coffee. Supporting this conclusion, it was recently reported that free valine, leucine and isoleucine are 2 to 3 times more abundant in Robusta green coffee than in Arabica (19).

Another opportunity to apply the same statistical approach was found during a storage test. In that experiment, the evolution of coffee volatile composition was tentatively characterized on the basis of headspace profiles collected at various times over a four month period. The sample consisted of ground coffee packed in small aluminium capsules. Two series of capsules had been prepared, differing by the composition of the enclosed atmosphere: normal air composition (A), or reduced oxygen (B). The storage temperature was 20°C.

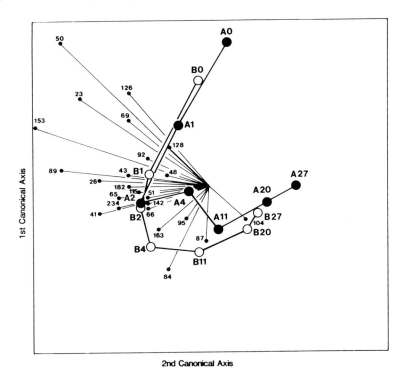

Fig. 2. Canonical plot showing the evolution of coffee volatile composition in ground coffee stored at 20°C under normal atmosphere (A) or reduced oxygen (B). Numbers in coffee labels indicate the storage period in weeks.

As before, the set of headspace profiles was sucessively analysed by SDA and CA. The result of this treatment is shown in Fig. 2. In this projection, the continuous evolution of the volatile composition appeared quite clearly. From the shape of the coffee trajectories, it could be inferred that at least two distinct processes had occurred during the considered period. The information was too limited to determine the nature of these processes. However, the very similar time dependence characterizing most components seemed more likely to result from diffusion processes than from chemical reactions. This conclusion was apparently supported by the absence of any significant difference between samples A and B.

Correlation of Sensory and Chemical Data

The last application of multivariate statistics to be reported here was an attempt to correlate coffee sensory characteristics to its volatile composition. This experiment was based on the 15 coffees defined above. The sensory evaluation consisted in a quantitative descriptive analysis (QDA) in which 12 descriptors of coffee aroma and flavour were to be scored on a seven-point scale (Table 1). A panel of 12 tasters had been trained to perform this task. The outcome was a set of sensory profiles which formed the counterpart to the already available headspace profiles.

Table 1: Coffee Sensory Descriptors

Aroma	Aroma and Flavour	
1 Intensity	4 Purity	9 Acidity
2 Richness	5 Strength	10 Astringency
3 Fineness	6 Balance	11 Body
	7 Roasted	12 Appreciation
	8 Bitterness	

As a preliminary to the actual correlation study, the QDA data were analysed by CA. The result of this analysis is illustrated in Fig. 3. As in Fig. 1, it could be established from the trajectories of the three coffee varieties that roasting was represented in the canonical space by the first axis. Coffee basic sensory characters like bitterness, acidity, roasted note, etc. were closely correlated to this roasting axis. On the other hand, the remaining descriptors formed a tight bundle pointing along the second axis. This clearly designated it as the quality axis. The existence of an optimum roasting for each coffee variety was confirmed by the shape of the trajectories. Each of them formed a half-circle around the quality axis.

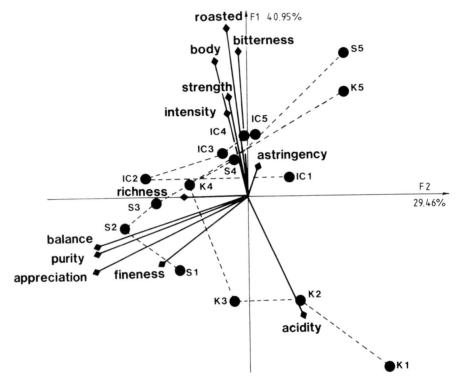

Fig. 3. Canonical analysis of the QDA profiles of the 15 coffees shown in Fig. 1.

A first approach to determine existing correlation between coffee sensory and chemical variables was based on the Procrustes analysis (16). By this procedure, different canonical configurations describing the same set of products are tentatively superposed by topological transformations in the same hyperspace. There, the correlations between the different kinds of variables can be determined from the angle formed by the corresponding vectors. Fig. 4 shows the result of this treatment when applied to the headspace and sensory data configurations. For the sake of clarity, the product configurations have been left out and only the two sets of vectors are represented.

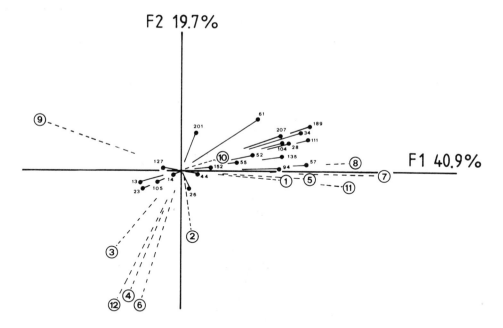

Fig. 4. Procrustes analysis of coffee headspace and QDA profiles (Fig. 1 and 3). Vectors representing headspace components (—) or sensory descriptors (---) are merged in the same canonical space.

In this projection, the overlapping area of the sensory and chemical analysis was immediately apparent. Along the roasting axis, the basic sensory characters were found to be correlated to a fairly large number of components. Conversely, none of the quality descriptors seemed to be directly related to any component. This latter observation corroborates the fact that the subjective evaluation of coffee quality is based on various appreciation elements which cannot be determined by just a few components. On the other hand, headspace profiles could be used to determine the basic sensory characters of a coffee.

As a complement to the Procrustes analysis, an attempt was made to find regression functions relating the two kinds of variables. Two different mathematical models were tested:

Linear model $\qquad S = a_o + a_1 X_1 + a_2 X_2 + \ldots$

Logarithmic model $\quad S = a'_o + a'_1 \log(X_1) + a'_2 \log(X_2) + \ldots$

where S is the sensory score for a given descriptor, and X_i the abundance of the ith headspace component. The selection of the appropriate variables and the determination of coefficients a_i and a'_i were achieved by means of the stepwise regression analysis (SRA). The components were entered one by one into the models until no significant improvement could be brought to the correlation. As shown in Fig. 5 in some instances a good correlation was already obtained with the first selected component. As a rule, slightly better correlations were achieved with the logarithmic model.

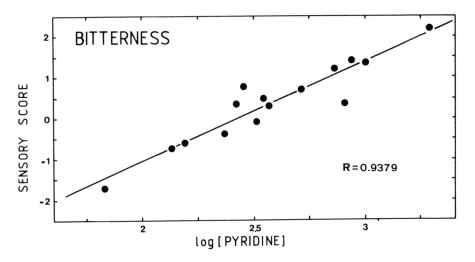

Fig. 5. Plot of bitterness average score for each of the 15 coffees versus the logarithm of pyridine abundance in the headspace profiles.

Table 2: Regression model for coffee roasted note

$S_{Roasted} = -10.90 + 4.95 \log [\text{Ethanal}] - 1.89 \log [\text{Isobutanal}] +$
$1.83 \log [\text{Butenal}] + 2.41 \log [\text{Pyridine}] - 3.75 \log [\text{Peak } 127]$

Multiple R $= 0.9900$
F-ratio $= 88.60$ (significant at 0.1% confidence level)
Standard Error of Estimation $= 0.17$

The detailed regression model obtained for coffee roasted-note is reported in Table 2. Highly significant functions were also found for the other basic sensory characters. On the other hand, the functions obtained for the quality descriptors clearly were of much lower significance. The validity of the various models was tested by comparing the experimental sensory scores with the values calculated by means of these regression functions. The result of this comparison for the roasted-note is shown in Fig. 6. As expected, for the 6 basic characters the experimental and calculated scores were in good agreement. Conversely, the regression functions obtained for the quality descriptors appeared to have practically no predictive power.

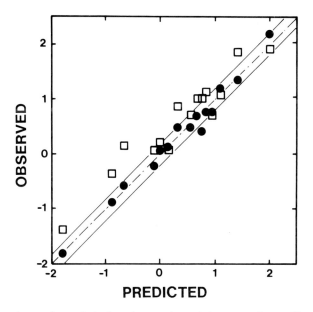

Fig. 6. Comparison of predicted and experimental scores for coffee roasted note. (● QDA 1st replicate, □ 2nd replicate).

Conclusion

Using a purely statistical approach, it was possible to extract from the volatile composition of a product like coffee various pieces of information which could serve to characterize this product. Whenever possible, the reliability of this approach was checked by confronting the conclusions based on the statistical analyses with otherwise available data. In most instances, a good agreement was observed.

Among the different applications which have been included in this study, the best documented is the discrimination of samples belonging to different categories. According to the recent literature, this application has been successfully tested on widely different products. The results presented here for the discrimination of coffee roasting and/or variety confirmed that this product did not stand as an exception.

Another potential use of the statistical approach has been found in the study of the processes which occur in such complex products as foods. As shown in this study, a technique like canonical analysis can provide in a single treatment a complete overview of the influence of the product parameters on its composition. This approach could be an alternative to studies based on simplified model systems which do not represent the complete reality.

The results of our attempt to correlate sensory and chemical data are encouraging. Hopefully similar studies will be repeated on coffee or other food and will confirm the validity of the chosen approach. However, this application still appears as a very complex one. Critical factors are the reliability and relevance of the sensory data. On the other hand, in order to obtain realistic correlation models, it will be necessary to take into account synergism or any other interaction effect between the components.

Acknowledgement

We express our thanks to H. Rahim for his contribution in the development and use of computer programs for multivariate statistics. We are also grateful to L. Vuataz for his help in the interpretation of the results of the statistical treatments.

References

1. Powers, J.J., Keith, E.S.: J. Food Sci. 33, 207 (1968).
2. Young, L.L., Bargmann, R.E., Powers, J.J.: J. Food Sci. 35, 219 (1970).
3. Persson, T., von Sydow, E., Akersson, C.: J. Food Sci. 38, 682 (1973).
4. Persson, T., von Sydow, E.,: J. Food Sci. 39, 537 (1974).

5. Lindsay, R.C.: Flavour Quality. Objective Measurement, R. C. Scanlan ed., ACS Symposium Series, Washington (1977).

6. Aishima, T.: Agric. Biol. Chem. 43, 1711, 1905, 1935 (1979).

7. Aishima, T.: J. Food Sci. 47, 1562 (1982).

8. Kwen, W.O., Kowalski, B.R.: J. Agric. Food Chem. 28, 356 (1980).

9. Noble, A.C. Flath, R.A., Forrey, R.R.: J. Agric. Food Chem. 28, 346 (1980).

10. Schreier, P., Reiner, L.: J. Sci. Food Agric. 30, 319 (1977).

11. Snygg, B.G., Andersson, J.F., Krall, C.A., Ställman, U.M., Akersson, C.: Appl. Envir. Microbiol. 38, 1081 (1977).

12. Cole, R.A., Phelps, K.: J. Sci. Food Agric. 30, 669 (1979).

13. Biggers, R.E., Hilton, J.J., Gianturco, M.A.: J. Chromat. Sci. 7, 453 (1969).

14. Ott, U., Liardon, R.: Flavour '81, P. Schreier ed., Walter de Gruyter, Berlin (1981).

15. Biomedical Computer Programs P-Series, W.J. Dixon and M.B. Brown eds., University of California Press, Berkeley (1977).

16. Gower, J.C.: Psychometrika 40, 33 (1975).

17. Vuataz, L.: Nestlé Research News 1976/77, 57 (1977).

18. Liardon, R., Ott, U.: Food Sci. Technol, in press.

19. Tressl, R.: 10th Colloquium on Coffee Chemistry, Salvador, Brazil (1982).

AUTHOR INDEX

462

Rapp, A. 121
Reineccius, G.A. 19
Roukeria, S. 93

Schamp, N. 381
Schlegelmilch, F. 93
Schomburg, G. 121
Schreier, P. 293
Schultze, W. 307

Shibamoto, R. 233
Sköries, H. 49
Sugisawa, H. 3,357
Suihko, M. 433

Takeoka, G. 63
Tressl, R 323

Weeke, T. 121
Willaert, G. 381

SUBJECT INDEX

SIEMENS

New horizons in GC

Run two analyses simultaneously

Take the twin-oven version of the SiCHROMAT®, and you have two gas chromatographs for little more than the price of one – a laboratory tool that does far more than just double your analytical capabilities.

With it, you can

- run two working temperatures in the same device, in three different ways:
  - both chambers equally or differently isothermic
  - one chamber with a temperature program, the other isothermic
  - both chambers with a temperature program

- add a new dimension to column switching, e.g. using columns of different polarities at different temperatures for higher resolution, shorter analysis times, thus increasing detection limits

- double the column space by taking out the partition, for extra-long columns and a total of 8 detector and injector connections

- carry out two independent routine examinations at the same time.

For a detailed description of the advantages of the new Siemens SiCHROMAT, ask for our 32-page brochure now.

with the twin-oven model of the Siemens SiCHROMAT

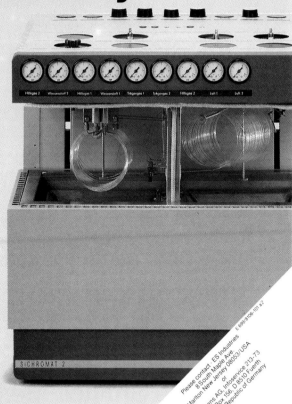